HEART OF THE COAST

HEART OF THE
OF THE

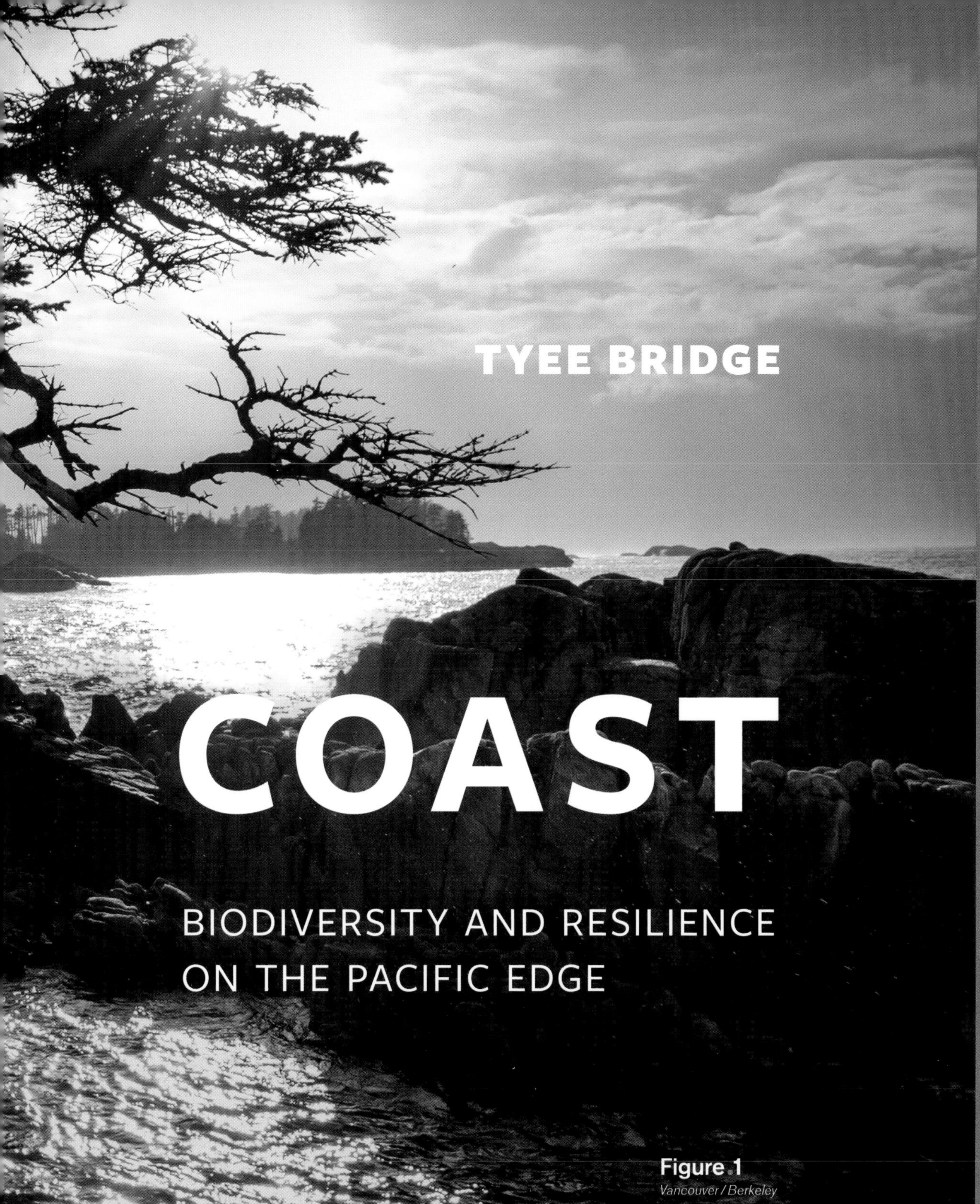

TYEE BRIDGE

COAST

BIODIVERSITY AND RESILIENCE
ON THE PACIFIC EDGE

Figure 1
Vancouver / Berkeley

This book is dedicated to the scientists, researchers, activists, and advocates from all nations who are working to understand and protect the miraculous diversity of life on Earth.

And to my son, Jonah, who like the rest of his generation will inherit the future we leave him.

Cataloguing data are available from Library and Archives Canada
ISBN 978-1-77327-115-6 (hbk.)

Design by Naomi MacDougall
Photography by Grant Callegari, except where otherwise noted. Additional photography by: Dean Azim: 118 *left*, 165 *top*; Shanna Baker: 129 *top*; Steve Beffort: 30; Mandy Bemis: 48, 49 *top right*, 132 *top right*, 133 *top right*; David O. Brown: 70; Jenn Burt: 10, 96 *top*; Stewart Butler: 24, 27 *top*; Tavish Campbell: 29, 33, 38, 40, 41, 42, 46, 75 *top*, *bottom*, 101, 115 *bottom*, 119, 138, 141, 158, 164, 180; Alex Cebulski: 36; Amy M. Chan: 55; Emmett Duffy: 100 *left*; Chris Ernst: 146 *right*, 150 all except *top right*; Bill Floyd: 28 *top*, 31 *top*, 32; Erin Foster: 7 *bottom*, 91 *top*; Alyssa Gehman: 62; Ian Giesbrecht: 27 *bottom*, 28 *bottom*, 31 *bottom*, 35; Kyle Hall: 73 *bottom*; Leslie Harris: 132 *top left*; Julian Heavyside: 140 *right*; Margot Hessing-Lewis: 78, 100 *right*; Kira Hoffman: 140 *left*; Maxwel Hohn: vii, 81 *right*, 104 *right*; Keith Holmes: 73 *top*, 84 *top*, 85, 142, 152 *top*; Nancy Knowlton: ix; Jeremy Koreski: 167 *bottom right*; Nick Irwin and Varsha Mathur: 131 *left*; Neil McDaniel: 61; Joanne McSporran: 14 *top*, 16, 20; Ryan Miller: 76, 81 *left*, 105 *top*, 106 *top*; P. Moore, Parks Canada: 97; NASA (Joshua Stevens/Jesse Allen): 114; Chris Neufeld: 72; Mark Ohman, Scripps Institution of Oceanography: 47; Angeleen Olson: 104 *left*; Gustav Paulay: 48, 49 *top right*, 132 *bottom left*, *bottom right*, 133 *top left*, *bottom left*; Katrina Pyne: 122 *top*; Luba Reshitnyk: 106 *bottom*; Amy Romer: 53; Kevin Schafer/Alamy: 60; Josh Silberg: 49 *top left*, 90 *top*, 118 *right*, 124 *top right*, 125, 130 *right*, 131 *right*, 133 *bottom right*, 134, 136 *top*, 149, 165 *bottom*; Kevin G. Smith/Alamy: 23; Brian Starzomski: 147 *top*; Tom Suchanek: 82; Sara Wickham: 146 *left*, 147 *bottom*; Mark Wunsch: 89 *left*; Kiliii Yuyan: 108, 109 all except *top left*.

Maps by Keith MacLachlan, 360GIS Solutions Inc.
SOURCES: Northeast Pacific coastal map, pp. x–xi: Ocean, Rivers, Lakes, and Shoreline polygon layers provided care of the University of Hawaii and NOAA (National Oceanic and Atmospheric Administration) and rights therein granted under the GNU license framework. Beringia and expansion routes map, p. 18: Estimated ice cover and Beringia exposed land informed by Bond 2019, "Paleodrainage map of Beringia" and correspondence with Jeffrey Bond, Yukon Geological Survey; PIDBA "Fluted Points 13,000 cal BP" map; Lesnek et al. 2018, "Deglaciation of the Pacific coastal corridor directly preceded the human colonization of the Americas"; Potter et al. 2018, "Current evidence allows multiple models for the peopling of the Americas"; Waters 2019, "Late Pleistocene exploration and settlement of the Americas by modern humans"; and Dyke 2004, "An outline of North American Deglaciation with emphasis on central and northern Canada." Ocean, Rivers, Lakes, and Shoreline polygon layers provided care of the University of Hawaii and NOAA and rights therein granted under the GNU license framework. Icefields to oceans graphic, p. 34: Based on O'Neel et al. 2015, "Icefield-to-ocean linkages across the northern Pacific coastal temperate rainforest ecosystem." Adams River sockeye migration map, p. 41: Lakes (Inland Water Bodies) Sources: Esri, Garmin International, Inc. (formerly DeLorme Publishing Company, Inc.). Ocean, Rivers, Lakes, and Shoreline polygon layers provided care of the University of Hawaii and NOAA and rights therein granted under the GNU license framework. Marine food webs graphic, p. 44: Based on Hakai Institute graphic "Simplified Food Web in the Northern Salish Sea" by Josh Silberg.

Illustrations by Emily Damstra

Copy editing by Lana Okerlund
Proofreading by Lucy Kenward
Indexing by Audrey McClellan
Front cover photograph by Grant Callegari
Back cover photographs by Grant Callegari except bottom right by Tavish Campbell

Printed and bound in Canada by Friesens
Distributed internationally by Publishers Group West

Figure 1 Publishing Inc.
Vancouver BC Canada
www.figure1publishing.com

CONTENTS

AUTHOR'S PREFACE

THIS BOOK WAS born from an email. Eric Peterson and I were corresponding about a project involving the University of British Columbia, and he mused: "Maybe somebody should help us write a book about the Hakai odyssey at some point."

It was early 2019, and by then I had written a few feature articles for *Hakai Magazine*. Like the Hakai Institute itself, the online magazine is funded by Eric and his wife, Christina Munck, through the Tula Foundation. The articles on coastal science and society that editor Jude Isabella manages to wrangle for it are excellent—creeling for langoustine "lobsters" in Northern Ireland, the mystery of a sunken Civil War submarine, and a parade of features by great writers like Andrew Nikiforuk, Amorina Kingdon, Christopher Pollon, and J.B. MacKinnon. The site's high standards have won it more than a few digital publishing awards. While the magazine sometimes interviews researchers from the Hakai Institute, at the time of my email exchange with Eric I had no real idea of the work that the institute does. So I looked into it.

I was impressed, and a little amazed. Coastal archaeology, biological and physical oceanography, marine food-web research, "bioblitz" surveys, kelp forest ecology, sand-dune tracking with laser scanners: these are only part of what Hakai does, most of it coordinated through research facilities on Calvert Island and Quadra Island.

Scientists and ecological knowledge-holders of all kinds have too often been stifled and stymied in recent decades. To see that for the past 10 years an independent institute has been carrying out challenging (and expensive) research projects—in collaboration with science hubs like the Smithsonian Institution and a wide variety of university departments—was a welcome surprise.

I jumped on Eric's idea of a book and badgered him to meet with me and publisher Chris Labonté. The conversations that ensued resulted in this book, which attempts to describe a few dynamic areas of Hakai Institute research and some of the stories and discoveries that have come out of them.

In the process, dozens of scientists shared their knowledge with me, as well as their love and enthusiasm for the subjects they study, from zooplankton to sea otters to glacial ice. Thanks to these researchers, writing *Heart of the Coast* has been fascinating and inspiring. I hope some of the curiosity, concern, and passion that drive their work comes through in these pages.

TB
New Westminster, BC

From glaciers to ocean, coastal British Columbia offers rich and diverse habitats for scientific study.

FOREWORD

I AM A coral reef ecologist, so most of my time has been spent far from the chilly waters of the northeast Pacific. But the tropics are not where I started. I first learned how to collect scientific data along the shores of the Strait of Georgia, whose waters are central to the stories upon which this book is built. I got my scientific diving credentials just to the south, off the coast of central California, where many of the creatures featured here are also found. Reading *Heart of the Coast* was like opening an old scrapbook, reawakening memories of the roots of my life's work. The waters may be cold, but the memories are warm.

It can't be denied, however: the northeast Pacific is a challenging place to do scientific research, as we humans lack the blubber of the seals or the fur coats of the sea otters that hunt here. The seawater that runs through the marine laboratories of the region is so frigid that your hands ache within seconds of immersing them. The thickest wet suits are not enough to keep warm, and even with dry suits, scuba diving is a physiologically draining experience. Extreme tides and strong waves create currents and surges that make it difficult to do the most elementary tasks, or even simply stay in place. It is more than difficult; it is dangerous. People have died trying to understand what makes this part of the ocean tick.

Despite these challenges, the scientific payoffs to studying the northeast Pacific coast have been enormous. Indeed, much of what students learn and professors teach in marine biology classes today was gleaned from these efforts. *Heart of the Coast* opens up this world to anyone who delves into its pages. We learn of the complex relationships linking sea stars, sea urchins, kelp, and sea otters, as well as the human-caused disruption of these connections driven by our once insatiable demand for sea otter pelts. It is also a warning of what might lie ahead, delivered by the scientists who discovered and then documented in heartbreaking detail how once abundant sea stars have vanished. They were the victims of a still poorly understood disease far more contagious and deadly than the coronavirus, one whose catastrophic impacts were fueled by climate change.

Perhaps most importantly, this is a deeply human story, and not just one of the despair that comes from counting thousands of dying and dissolving sea stars. It is also the story of the archaeologists who discovered footprints left along the shore by humans who traveled the "kelp highway" 13,000 years ago. It is the story of the bioblitzers enchanted by each new worm, crab, fish, and snail they discover during a three-week effort to record what lives in the region. It is the story of the insights that have come thanks to the generosity of philanthropists who have committed their resources to exploring, protecting, and communicating about the waters of the region. In the end, it is the story of a passionate commitment to place and the wisdom and joy that this brings. In that sense, despite all the concerns, it is a story of ocean optimism.

NANCY KNOWLTON
Sant Chair for Marine Science Emerita, Smithsonian National Museum of Natural History, and author of *Citizens of the Sea*

BERING SEA

Aleutian Islands

CENTRAL COAST

BRITISH COLUMBIA

Great Bear Rainforest

Haida Gwaii

HECATE STRAIT

Bella Bella ●

● Bella Coola

• Koeye

Calvert Island ●

QUEEN CHARLOTTE SOUND

Port Hardy ●

0   45   90   180 Kilometers

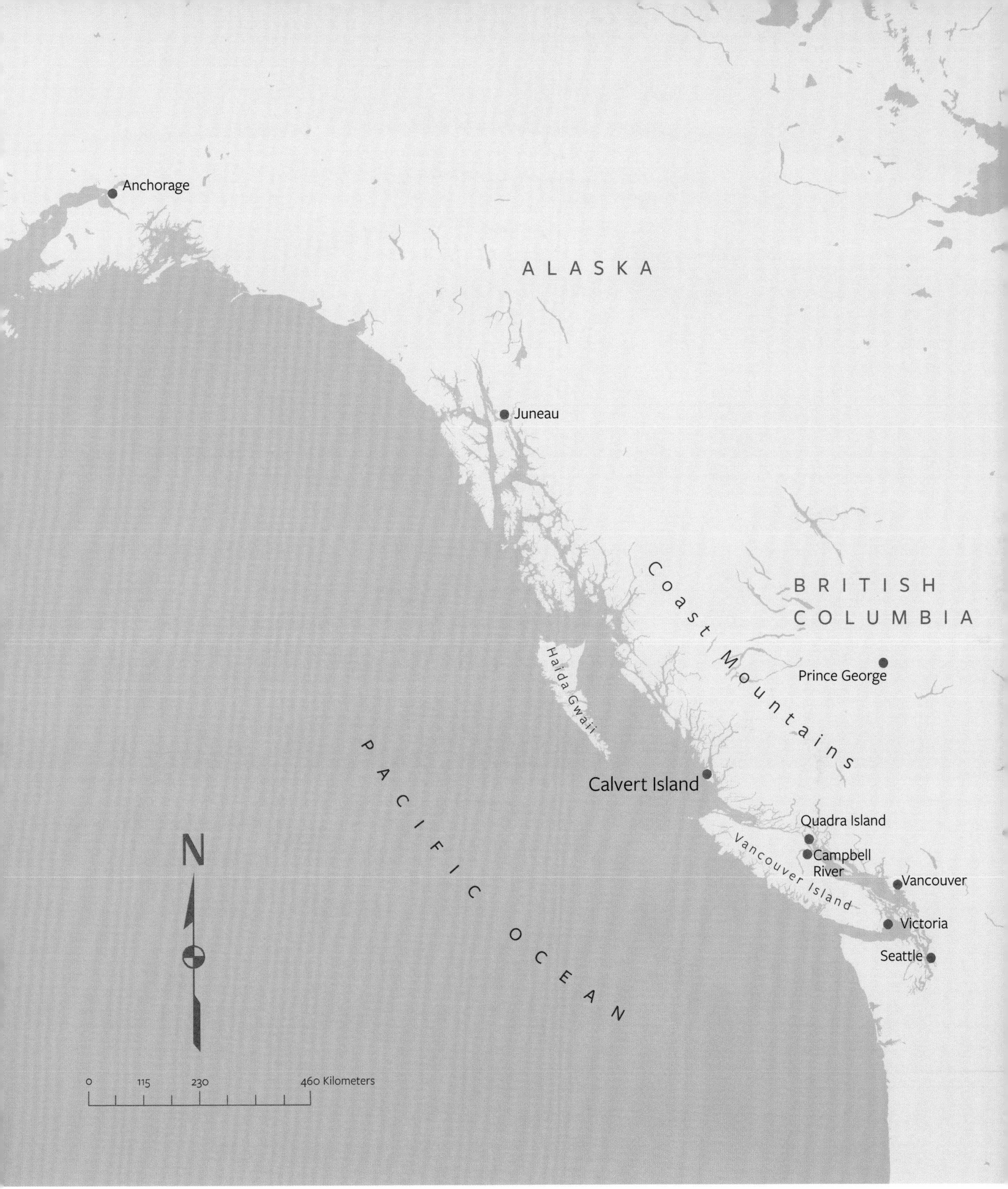

ALASKA

Anchorage

Juneau

Coast Mountains

BRITISH
COLUMBIA

Prince George

Haida Gwaii

PACIFIC OCEAN

Calvert Island

Quadra Island

Campbell
River

Vancouver Island

Vancouver

Victoria

Seattle

N

0   115   230   460 Kilometers

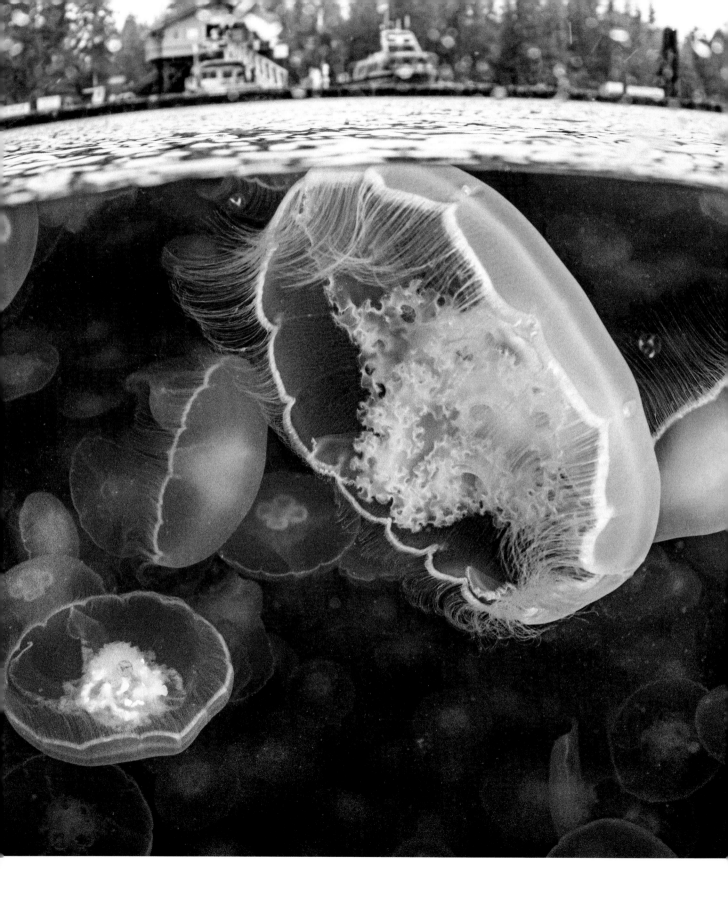

INTRODUCTION

THE ORIGINS OF THE HAKAI INSTITUTE

ERIC PETERSON Cofounder and President, Tula Foundation and Hakai Institute

FOR THE FIRST 30 or so years of my life, despite a few diversions, I had no doubt about what I wanted to do with my life. I wanted to be a scientist: to teach and do research in my own laboratory at a university. I started in mathematics, fell in love with genetics and molecular biology, and ended up as a neuroscientist. I began near home at the University of British Columbia, went to the University of Sussex for a PhD (where I met my future wife and Tula collaborator, Christina Munck), moved to Harvard for a postdoc, then landed a research faculty position in McGill University's biology department. Mission accomplished—or so I thought.

Setting up a new lab and research program at McGill was an interesting challenge, and teaching was a pleasure. But over the next few years, my enthusiasm at the prospect of a life in academia began to wane. My eye began to wander to the career sections of weekend newspapers.

In those years, the mid-1980s, these sections were filled with technology jobs. Many of these opportunities looked far more exciting than what I was doing—and they paid much more. So, in the biggest decision I'd ever made in my life to that point, I quit my university job, answered one of those ads in the newspaper, and threw myself into the tech sector.

I was drawn to manufacturing: automation, robotics, machine vision. Prior to my PhD, I had worked for a couple of years in power production and process control in a pulp mill, so I was at home on the factory floor. Bringing all the pieces together—instrumentation, software, toolmakers, engineers, production managers, workers—was an engaging challenge. Many projects I worked on were heroic failures.

As I found myself reenergized in my new career, I realized that I gravitate more toward engineering than science. Engineers build things, solve problems, move from one project to another. I enjoy that; I lack the enduring patience that it takes to sustain an independent scientific career. To my great surprise, I also found that I was actually pretty good at organizing, managing, and bringing together people smarter than I am to solve tough problems.

Serendipity brought me to the hospital sector. Hospitals, despite their focus on human health, are like factories in many ways. The principles that apply to advanced manufacturing—information management, automation, quality assurance, reduction of what is known as "work in progress"—all apply there as well. And nowhere was the need greater than in diagnostic imaging, where the explosive growth of computerized tomography (CT), magnetic resonance imaging (MRI), ultrasound, and other technologies was overwhelming hospitals' limited resources.

With a partner, I cofounded Mitra in 1990. Over the next decade, Mitra became a key player in the evolution of medical imaging: from light boxes and stacks of X-ray film to computer workstations, digital archives, and electronic health records. Mitra's success and sale led to the launch of the Tula Foundation at the end of 2001.

Models and mentors

Christina and I created the Tula Foundation as a way to support what we broadly describe as "innovations and solutions in the public interest." In selecting Tula programs, then as now, we seek topics that are compelling to us and that we think we can do well. In the past two decades, we have been able to support a wide variety of important programs. Along with the creation of the Hakai Institute, a brief list includes longstanding commitments to our rural healthcare program in Guatemala, which began in 2002; the University of Victoria's Environmental Law Centre; and independent media such as *Hakai Magazine*.

Over our first decade, we moved Tula from being a donor organization funding projects run by others to being an operational entity with its own programs. In hindsight, it seems inevitable that we would launch a significant science program on the BC coast, and that it would become our central and most ambitious effort. Christina and I both grew up near the ocean, and we both first opted for a career in biology. We'd already spent a few summers cruising the coast to Alaska and Haida Gwaii in our boat, so we knew the territory well.

For many years, we'd been funding land acquisition and stewardship on the coast while supporting individual scientists to do research on a limited scale. We often found it frustrating to watch gifted people working without adequate resources, local infrastructure, or a long-term plan. Our attempt at a solution began in 2009, when we purchased the property that would become Hakai's Calvert Island Ecological Observatory.

When I do anything, I look for a model to emulate. At Mitra my model was always Hewlett-Packard (HP)—the gold standard for tech companies in its day. HP was an engineers' company where the namesakes—just Bill and Dave to everyone—had created an open, innovative, collaborative culture. I had Dave Packard's *The HP Way* on my bookshelf.

Starting in the mid-1980s, Packard concentrated on nonprofit initiatives in marine science. The Monterey Bay Aquarium was one of his flagship endeavors. I was particularly interested in Packard's companion venture, the Monterey Bay Aquarium Research Institute (MBARI), which had quickly established a stellar reputation for oceanographic research and for developing deep-sea exploration technology. What had Packard's game plan been? Had he brought the "HP way" to marine science? Where were the lessons for me?

Fortunately, Packard's thinking was well documented. I could see that we were aligned on the big principles. If Hakai were to emulate some of Packard's successes, like MBARI, it would need a clear identity and a long-term vision. It would also need to be independent, particularly financially, and not subject to the politics and whims of research funding.

Packard, as always, stressed the importance of organizational culture in developing MBARI. He insisted that scientists, engineers, and operations staff work together in close collaboration. And like any good strategist, Packard had identified the advantages available to MBARI. He noted that Monterey Bay, with its steep drop-off, offered better access

Pruth Bay and the dock of the Calvert Island Ecological Observatory.

to the deep ocean than anywhere else on the Pacific coast of America; therefore, MBARI's niche would be technology to explore these depths.

From the start, Christina and I had a general sense that the Hakai Institute should be a center for science, education, and community engagement. It satisfied our three criteria: it was important; it was interesting to us; and it looked like something we could do well. We knew it would be a challenge, but we felt well prepared for the mission. However, we certainly didn't have a deep ocean canyon in our backyard at Hakai. What were our advantages?

There turned out to be several. Partly because of its remote location, Calvert Island has remarkably well-preserved biodiversity. The area is embedded in an extensive network of nature conservancies—marine and terrestrial—meaning that its future evolution would primarily be driven by natural forces rather than human development. No place on the planet is truly pristine, but the Central Coast location of our ecological observatory is relatively free from industrial development and invasive species.

The Central Coast is also First Nations' traditional territory, including that of the Heiltsuk, Kitasoo/Xai'xais, Nuxalk, and Wuikinuxv. In many areas of our work, there are natural opportunities to incorporate local and traditional ecological knowledge with formal science to the benefit of all parties. We have strived from the outset to reach out to our First Nations neighbors to ensure that our work respects the social context of the Central Coast.

At the outset, we had no specific plan in mind, apart from trying to fill the place with great science and education. From the start, we drew heavily on the university partnerships we had established through our previous work. Hakai seemed to me to be a natural laboratory, a possible reference site for monitoring biodiversity and climate change that would be of significance for the entire coast. Like an enthusiastic undergraduate, I tried this seemingly novel concept out on Ken Lertzman, a forest ecologist in Simon Fraser University's Faculty of Environment. Ken sighed and gave me a long reading list to educate me on where my ideas fit into established ecological thinking.

Ken introduced me to the Long Term Ecological Research (LTER) Network. Since 1980, the US National Science Foundation has funded a nationwide network of reference sites, supporting ecological discoveries that are related to the influences of long-term and large-scale phenomena. Ken's suggestion was bang on. Being Canadian, we couldn't formally join the LTER Network, but we sought and received great cooperation and were able to eventually adopt all the LTER tools, research strategies, and conceptual framework. We got all the gear, as it were, like a new franchisee joining an established chain.

Ken gave me a long list of research facilities that might serve as models, including La Selva in Costa Rica, operated by the Organization for Tropical Studies, and Barro Colorado Island in Panama, operated by the Smithsonian Institution. Through Ken and other colleagues, I also learned about other antecedents on the West Coast. These included the Partnership for Interdisciplinary Studies of Coastal Oceans (PISCO) and the work of Bob Paine—particularly his intensive long-term studies on Tatoosh Island off the tip of Washington State's Olympic Peninsula (see page 57).

TOP: Hakai and Simon Fraser University researchers seining for juvenile salmon in Johnstone Strait.

BOTTOM: This Ocean Networks Canada oceanography monitoring buoy in Baynes Sound provides data on acidification that can affect shellfish farms in the region.

TOP: Carolyn Prentice, foreground, and Brenna Collicutt processing urchins in the Marna Lab at Hakai's Quadra Island Ecological Observatory.

BOTTOM: The Marna Lab has an array of mesocosms—tanks that mimic ocean conditions and allow researchers to test the effects of altered pH, salinity, temperature, and other factors.

Ken also introduced me to the concept of ecological resilience. Resilience of social-ecological systems was the explicit topic of several of our early academic courses and symposia, and it continues to be an enduring theme of the Hakai Institute.

Catalyzing collaborations

Thanks to Ken and these models, we had the makings of a plan. We just had to put it into action. With the conclusion of our Rivers Inlet program—another source of inspiration, conducted in partnership with BC universities and the Wuikinuxv Nation—our energies as the newly formed Hakai Institute moved into some of the areas we are still focused on and passionate about.

This included a vibrant oceanography program, sampling for marine biogeochemical data; monitoring bog forests for their microbiological and chemical interaction with the coastal ocean; an archaeological program that continues to produce fascinating discoveries about early human habitation of the Central Coast; and studies of the incredibly rich intertidal areas around Calvert, including kelp and seagrass habitats.

We are comfortable working incrementally and opportunistically, starting with projects that are close at hand and easy to do with the partners we know, and expanding from there. Over time, this incremental approach has taken us to where we are today—with research ranging from the edge of the deep ocean to the coastal glaciers, and working with partners from the US Pacific Northwest to Alaska.

In 2014 we established a second ecological observatory, roughly 200 kilometers to the south on Quadra Island, at the northern tip of the Salish Sea. We focus there on local priorities, including juvenile salmon migration and the biological effects of ocean acidification. Laboratory facilities support experimentation and analysis to complement the observational science we conduct across the coast. Hakai also has centers in Campbell

TOP: West Beach on Calvert Island.

BOTTOM: Coastal gray wolf in the Goose Islands, northwest of Calvert Island.

Researchers gather around a campfire on Calvert Island's West Beach.

River and Victoria and is embedded in universities in Vancouver (at the University of British Columbia) and Prince George (at the University of Northern British Columbia).

My experience with Mitra informs the work we continue to do through the Hakai Institute. As with Mitra, we are integrated from local to global scale: Hakai enjoys great grassroots relationships with institutions and researchers in our own backyard as well as with renowned international bodies like the Smithsonian. The issues that the Hakai Institute tackles also share many qualities with those we dealt with at Mitra; they combine science, technology, sociology, and politics and require us to find common ground with diverse stakeholders. We love science in the spirit of exploration and discovery. We also want to see our work serve the public interest via a contribution to enlightened policy and resources management in coordination with governments and First Nations.

Throughout all these ventures, we have been happy to act as catalysts, bringing the elements together and fostering collaboration.

A place for breakthroughs

In the early days, I gave some thought to the mystique of the research station, the expedition, the oceanographic cruise—catalytic locales where scientists are thrust together at close quarters and where revolutionary scientific ideas can be hatched. These intense, creative cultures spring up in unlikely places. I'd been swept up in one as a graduate student, where the departmental tearoom—a profoundly influential part of scientific and academic life in the United Kingdom—was an intellectual crucible, fueled by the restless genius of our head, the legendary evolutionary biologist and geneticist John Maynard Smith.

I'd seen it in intensive courses, such as the one I attended as a post-doc at the Cold Spring Harbor Laboratory on Long Island, where I was force-fed neurophysiology for two frenetic weeks. How could we foster this creative intensity and chemistry at Hakai?

Our first step was to recruit scientific partners, drawing first on the university connections we'd made through our previous research support. Beyond that, we've taken a tearoom approach. On Calvert, this amounts to throwing people together so they can naturally interact, and offering some of the infrastructure that makes that possible: bunkhouse accommodations; a large open dining room with big tables where everyone eats buffet-style together; after-dinner talks; and recreational opportunities on beaches and trails.

It's working well for us. More than once, we've seen situations where two researchers strike up a conversation on various areas of shared interest after chatting in the dining room food line—and are shocked to learn they've been working in the same building at the same university for 10 years, but have never met before. Thanks also to the remote location, I think we've managed to create a collegial, egalitarian atmosphere, where the dishwasher and the member of Parliament both fly in on the same plane and are treated with the same respect.

Initially it was a bit of a challenge for this whole enterprise to be taken seriously as a bona fide scientific institution, and to establish what Packard called a "clear identity and long-term vision." I feel that we have been able to strike a great balance that serves all parties.

That said, this book is not intended to chronicle all the work we've done at Hakai over our first decade—nor can it give due credit and place to our partners, friends, and everyone at Hakai Institute who have made it possible. That task would require 10 books. Rather, we simply hope to showcase some stories and snapshots that have come out of our work, and offer readers some insights into the amazing biodiversity and resilience of the Central Coast and beyond.

1 DEEP TIME
FOOTPRINTS OF THE PAST

THIRTEEN THOUSAND YEARS ago, two adults and a child were walking on an island in what is now British Columbia's Central Coast. The trio were in an area above the high-tide line of a beach; they may have been foraging for berries or preparing food, or they may have just disembarked from a raft or boat. Whether or not they were a family—and whether their people lived on the island for generations or were recent arrivals—is uncertain. But the fact of their presence is not, because they unintentionally did something remarkable: they left more than two dozen footprints that would never be completely erased.

During the period that these three people were tramping the soft, clay-rich soil on what is now Calvert Island, the northeast Pacific coast looked very different than it does today. Earth was in the tail end of the last ice age, and ice sheets were still in the process of receding from the continent. Lower sea levels exposed areas of the continental shelf, connecting landforms that are now islands and merging the future Alaska and Russia into a landmass called Beringia—a vast area of low-growing vegetation known as a mammoth steppe, roamed by now extinct species such as mammoths, saiga antelope, scimitar cats, and giant short-faced bears. Vegetation was also different: it would be over 4,000 years before cedar trees, one of the iconic species of coastal temperate rainforests, would advance north past the 49th parallel.

Fast-forward to a rainy afternoon in the spring of 2014, when archaeologists at this same location were digging into the wet near-shore sand at low tide. The effort, one of dozens of exploratory "shovel tests" by archaeologist Duncan McLaren and his team, was part of the Hakai Ancient Landscapes Archaeology Project. A Hakai Institute research initiative run with the participation of the Heiltsuk and Wuikinuxv First Nations, the project looks for evidence of early human occupation on the Central Coast. McLaren, his colleague Daryl Fedje, and a crew of graduate students and field assistants were excavating carefully but quickly that day because conditions for discovery were less than ideal: the rain was relentless, the tide was encroaching, and their freshly dug pits seemed to be yielding the same unimpressive results as their predecessors.

Fedje nonetheless felt a cautious anticipation that something might be different this time. Like McLaren, Fedje is an archaeologist who has been exploring the history of the BC coast for over 30 years. "About fifty to sixty centimeters below the surface, I noticed a change from modern shell hash and sand and beach textile to what looked like an old paleo soil, an old soil surface," says Fedje. "And when I hit that, I encountered this strange depression."

There wasn't much to see in the subsurface clay, but it was enough to stop an archaeologist with a keen eye. "It looked like it could be some kind of footprint. There was a kind of pattern recognition there, and I thought it might be a print of a bear or maybe a person—though it could have just as well been a place where a tree had fallen and made a kind of dimpled depression."

Fedje called McLaren over to look at the pattern pressed into the light brown clay at the bottom of the pit. They agreed it was an interesting find, but neither was sure exactly what they were looking at. The hole was deep, the light was dim, and rain continued pelting down as the seawater inched closer. They took a photograph and extracted some wood samples from the impression. Then they filled in the hole to protect their find from the incoming tide.

TOP: Daryl Fedje, left, and Duncan McLaren, right, in the excavation area, surrounded by other members of the archaeology team.

BOTTOM: Aerial view of the bays near the footprint dig site on Calvert Island. Vegetation would have looked very different here 13,000 years ago, but then as now the area would have been ideal for harvesting clams and other food.

TOP: A researcher holding up a find from the site. Small artifacts like this one tend to be easier to extract than footprints.

BOTTOM: Duncan McLaren examining the excavation.

Cycles of ice

Calvert Island is located in Queen Charlotte Sound between the southern tip of Haida Gwaii and the north end of Vancouver Island. This Central Coast area is the traditional territory of the Heiltsuk, Wuikinuxv, and Nuxalk First Nations. When people were pressing their feet into this soil 13,000 years ago—only a tick of Earth's geological clock, which goes back 4.6 billion years—Earth was near the conclusion of a long ice age. It was one of many ice ages that occurred in the current geologic period called the Quaternary.

The Quaternary began 2.6 million years ago. It is divided into two epochs: the Pleistocene, which lasted until 11,700 years ago; and the current epoch, the Holocene, which began at that time and continues today. Advancing and retreating ice has been a constant theme of the Quaternary. Repeated ice-age cycles have seen ice sheets (also known as continental glaciers) up to three kilometers thick move south from the Arctic to cover large portions of continents in the Northern Hemisphere, then shrink back to higher latitudes.

Earth's most recent ice age began about 110,000 years ago. At its peak, between 26,000 and 19,000 years ago—known as the Last Glacial Maximum—much of what is now Canada and part of the northern United States were blanketed by a thick cover of ice. Lands in the south of modern Alaska and almost all of British Columbia were blanketed by the Cordilleran ice sheet, which covered the area west of the Rocky Mountains and reached into present-day Washington State, Idaho, and Montana; this bordered the Laurentide ice sheet, which extended east from the Rocky Mountains across Canada and as far south as Ohio. By the end of the ice age 11,700 years ago, the ice sheets had effectively disappeared.

The last ice age had a significant effect on oceans and coastlines around the world. Because of the amount of water bound up in massive ice sheets, at the Last Glacial Maximum, global sea level was about 120 meters lower on average than it is today. Due to a variety of geological factors, however, ice-age sea levels in areas of the northeast Pacific coast varied wildly—ranging from 150 meters lower than the present day in Haida Gwaii to as much as 200 meters higher in British Columbia's Fraser Valley. One factor was the ice sheets' incredible weight, enough in some places to press the earth's crust hundreds of feet down into the upper mantle (a phenomenon known as isostatic depression), which in turn caused other land areas to bulge upward.

A question of sea level

McLaren and his team members began meeting with the Hakai Institute in 2010 to discuss possible avenues for research. Understanding the history of local sea levels was a critical first step.

"We were quite interested in the region from an archaeological perspective," says McLaren. "Based on our knowledge from other parts of the coast, we figured that there was a really good chance that sea level had been fairly stable in and around the Hakai Passage region for the last twelve to thirteen thousand years."

Before they could break ground in search of early human occupation, McLaren's team would have to confirm their theory. Two years of research later—testing sediment cores taken from ponds and lagoons, analyzing deposits of single-celled algae called diatoms, and using other methods—they found their hunch was right. Amidst an ice-age landscape of extreme variation, sea level on Calvert Island had remained relatively unchanged since the last ice age: only three or four meters lower than it is today.

Then began the shovel tests that revealed an apparent footprint. It took six months before the samples they'd taken from the foot-like impression would be dated in separate tests. (The first sample got lost for several months on its way to a California lab.) When they at last received the results, the data were promising: the tests placed both of the wood samples at around 13,000 years old.

LEFT: Daryl Fedje, right, and a team member extracting sediment cores in a Quadra Island bog.

RIGHT: Sampling the contents of a sediment core.

This gave McLaren and Fedje a reason to bring their shovels back to Calvert Island. If the impressions were made by humans, they would be the oldest human footprints ever found in North America to date, and one of the earliest known traces of human habitation in British Columbia.

Beringia and the "kelp highway"

Even during the Last Glacial Maximum when ice cover was at its peak, many areas on the northeast Pacific coast were untouched by ice. These refugia—places in the Northern Hemisphere where life could survive inhospitable ice-age conditions—offered thriving habitats for plants and animals. They could be found on the northeast Pacific coastal margin from southeast Alaska south to Haida Gwaii, Hakai Passage, and Vancouver Island.

Beringia, the largest ice-age refuge, stretched west from the Lena River in Siberia to the Mackenzie River in the Canadian Northwest Territories, encompassing Alaska and Yukon. As a whole, Beringia was home to a wide variety of now extinct species of land animals, as well as humans. Researchers theorize that humans gradually moved out of eastern Beringia at some point during the last ice age, their descendants moving westward and southward into North America over millennia. Beringia is something like the mythical land of Atlantis: as of about 12,000 years ago, much of it was covered by the ocean as sea levels rose. This now submerged "land bridge"—actually an encompassing area that at its peak was likely over a thousand kilometers wide—plays an important role in ideas of how the first people came to live in the Americas.

North and South America were the last continents on Earth to begin to host populations of *Homo sapiens*. How humans came to be here has been, and continues to be, a subject of intense debate. For many decades, the dominant theory, called the "Clovis first" hypothesis, held that the first peoples to inhabit North America gradually moved overland from Siberia via the Beringian land bridge about 13,500 years ago—then expanded south through present-day Canada via an ice-free corridor between the Cordilleran and Laurentide ice sheets. Researchers called them the Clovis people after a site in Clovis, New Mexico, where archaeologists in the 1930s found 13,000-year-old flint projectile heads.

However, while it is certain that humans did use this corridor east of the Rocky Mountains at some point during the latter part of the last ice age, more recent excavations have made it appear unlikely that this was

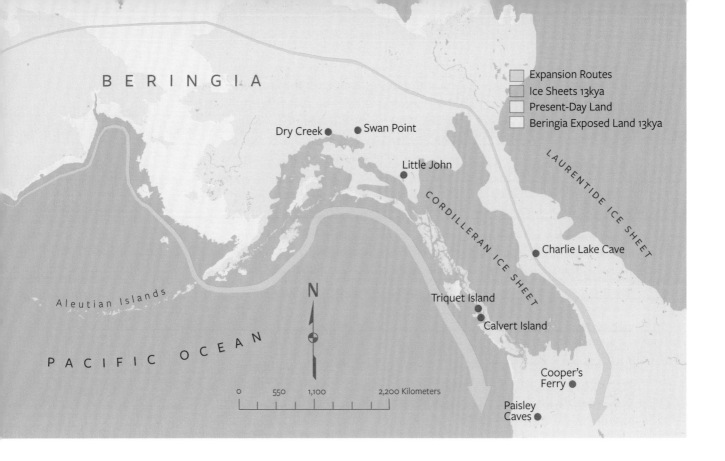

B E R I N G I A

Expansion Routes
Ice Sheets 13kya
Present-Day Land
Beringia Exposed Land 13kya

Dry Creek ● ● Swan Point

Little John ●

LAURENTIDE ICE SHEET

Charlie Lake Cave ●

CORDILLERAN ICE SHEET

Aleutian Islands

N

Triquet Island ●
Calvert Island ●

PACIFIC OCEAN

Cooper's
Ferry ●

0   550   1,100   2,200 Kilometers

Paisley
Caves ●

ABOVE: Beringia was a vast region stretching from Siberia to Yukon, an ice-age refuge that offered habitat for many now extinct animals. This map shows the encompassing land that, 13,000 years ago (known as 13kya), was still above sea level and connected what is now modern Asia to North America. Some of the significant archaeological sites relating to early human occupation are noted on the map.

Orange arrows indicate two potential routes of human territorial expansion from Asia at that time: the coastal route and the land route between the ice sheets. Current evidence indicates that a route between the ice sheets was not likely passable by humans until sometime after 14kya; archaeological finds indicating earlier human occupation—such as at Cooper's Ferry in Idaho, estimated to be 15.5kya—therefore lead many researchers to favor a coastal route as the way humans were first able to expand into lands south of Beringia.

RIGHT: Bone and stone artifacts found by archaeologists on Calvert Island.

how the first people arrived south of the ice sheets. Researchers believe the corridor was not open until between 15,000 and 14,000 years ago. Weapons, wooden tools, and stone artifacts have been found at the Monte Verde site in southern Chile that are dated to approximately 14,200 years ago. The journey from the ice-free corridor to southern Chile is a long one, and to populate that area, humans would have to have made their way south of the ice sheets a considerable amount of time prior to that period. At the Cooper's Ferry site in Idaho, archaeologists have found stone weapon points made between 16,500 and 15,300 years ago. And at the Gault site in central Texas, they have found projectile points dated as being at least 16,000 years old. Some are possibly over 20,000 years old.

Taken together, these findings have led many researchers to believe that a likely pathway for the first human settlement of the Americas was not between the ice sheets, but via the western coastline. By about 16,000 years ago, much of the outer coast of present-day Alaska and British Columbia was ice-free. Watercraft would have allowed people to colonize many islands and coastal areas over the ensuing millennia, settling areas from Alaska to Tierra del Fuego as they followed the abundant food resources and calmer waters of what has been called the kelp highway.

Following the footsteps

Compared to paleontological digs that find 60-million-year-old *Tyrannosaurus rex* bones, footprints 13,000 years old may not seem all that ancient: 520 generations, approximately. But they precede the building of the Giza pyramids by almost 9,000 years, and finding clues to the ways humans lived at that time takes careful work and preparation. When McLaren and his team were finally able to return to enlarge the excavation on northern Calvert Island, their long wait was rewarded: the unmistakable pattern of *Homo sapiens* footprints.

That's the kind of moment, says Fedje, when the hairs go up on the back of your neck. "We spend a lot of time looking for bits of chipped rock and animal bones and little things like that. We very seldom get the opportunity to see where somebody has been walking, and there's a larger person and a little person. That connects so much more viscerally with the human condition."

Humans take over 200 million steps in their lives, but finding intact footprints is a rarity—even those made in an ancient world devoid of asphalt and concrete. Whatever the optimal conditions are for making

long-lasting footprints, they were apparently present on Calvert Island 13,000 years ago: further surveying ultimately revealed a total of 29 identifiable prints.

"It turns out under very specific conditions, coastal regions can be ideal for preserving tracks," McLaren told an interviewer from *Hakai Magazine*. "Coasts have a lot of soft sediment where tracks or footprints are easily made or left. Then wind or wave action fills them with sand or silt."

The Calvert Island footprints are not old enough to offer conclusive evidence, but they fit with and support a history of long coastal occupation—as well as of the gradual movement of ice-age peoples southward along the coastal margin.

There is some nuance here, however, as the distinction between occupation and migration is less clear than it might appear. When archaeologists describe the peopling of the Americas with words like "migration" and "journey," it's easy to ascribe to those ancestral humans an intention that they likely didn't have. From a 21st-century standpoint, we may see a trajectory of people moving out of Beringia that culminates in human habitation at the tip of South America, but the people who ranged along the coast of the Americas had no final destination in mind. As much as they may support a theory of ice-age humans moving into the Americas via the coast, sites explored by the Hakai Ancient Landscapes Archaeology Project equally indicate a history of settlement.

"We talk in archaeology a lot about migration, and that seems to be a primary focus of what a lot of people spend their whole careers looking at," says McLaren. "But we're also quite interested in the long-term history of occupation of the place in and of itself. There is really good evidence that once people arrived in the Central Coast, they stuck around."

Historically, the area has been a very productive habitat, teeming with marine mammals like seals and sea lions, as well as rockfish, salmon, clams, and shellfish. Food and resources were relatively stable and easy to harvest. Given all the resources that were right in front of people, says Fedje, the area would have been a very comfortable place for people to live for thousands of years.

"So often, we have the story of human migration right into the Americas, as if people are a bunch of caribou or Canada geese," he says. "But no, these footprints show people using that particular bit of land and hanging out and walking and doing things. We don't know for sure because we have so few bits of information to put the story together. But in my mind, this is showing that people had become part of that landscape—that this was their home and they weren't bent on migrating, say, to Texas."

The kelp highway, then, is only a route of migration when seen in hindsight. Most peoples during the ice age and in the millennia that followed were likely content to call one part of it home for long periods. Some of those who lived in an area may have chosen to move elsewhere for any

Triquet Island, where Hakai's Ancient Landscapes project archaeologists found a cache of stone tools over 13,700 years old.

ABOVE: Artifacts found at another dig only a few meters away. Top, a remnant of a bone comb, likely 4,000 years old. Bottom, part of a harpoon toggle, approximately 2,000 years old.

RIGHT: Archaeologist Alisha Gauvreau at the footprint site on Calvert Island.

number of reasons—disaster, lack of food, social breakdown, or simply curiosity. When they did, they didn't necessarily always move south. People, says McLaren, could always make a choice. "They could say, 'Okay, we're going to stay put because things are pretty good here,' or they could move to the south or, conversely, move back northward."

One of the most striking confirmations of long occupation is an archaeological discovery at Triquet Island—a small island about 15 kilometers northwest of Calvert—made in 2016. Also part of Hakai's Ancient Landscapes project, the excavation on Triquet by McLaren, Fedje, and University of Victoria PhD candidate Alisha Gauvreau found a cache of stone tools associated with a hearth feature dated to between 14,000 and 13,700 years ago. Like the Calvert footprints, the Triquet artifacts confirmed for local Indigenous people their traditional history of long occupation of the area.

"Heiltsuk oral history talks of a strip of land in that area where the excavation took place," Heiltsuk member William Dúqvȧísḷa Housty told a news reporter when asked about the Triquet excavation. "It was a place that never froze during the ice age and it was a place where our ancestors flocked to for survival."

ANCIENT WATERCRAFT

In recent years, researchers have begun to over-turn the idea that people did not begin using boats for transportation until long after the ice age. Archaeological evidence from Australia, Southeast Asia, Okinawa, and other locations suggests that humans may have been using rafts and boats to cross significant distances for 50,000 years; research in the Philippines indicates that early humans related to ancestral *Homo erectus* were traveling to islands by boat as much as 700,000 years ago. The oldest watercraft yet discovered is the Pesse canoe, a 10,000-year-old dugout made from a Scotch pine, found in the Netherlands in 1955.

What sort of boats ancient humans in the Pacific would have used is unknown, as the archaeological record does not yet show any examples. Types of rafts, canoes, kayaks, and skin boats similar to Inuit umiaks (boats made from animal skin stretched over a wooden frame) are possible. ■

Iñupiaq seal-skin boat on the shore ice off Barrow, Alaska.

Shallower layers of the Triquet site also showed evidence of a lasting human presence.

"What we've found on Triquet is this beautifully preserved historical archive of repeated human occupation," says Gauvreau. "It shows that it really is one of those places where people have lived for millennia and remained connected to."

The Ancient Landscapes project has revealed a handful of sites on the outer coastal islands of British Columbia that show repeated occupation over 5,000 years, up to as much as 10,000 years. "And now, on Triquet Island, we see this very, very long-lasting, undisturbed record of repeated occupation that goes back into deep time, for at least fourteen thousand years," says Gauvreau. "That's what makes Triquet special; it is one of the oldest and longest-occupied sites currently known on the coast. We know there are other sites like that out there, but more work is needed."

2 FROM ICE FIELDS TO OCEANS

GLACIERS AND BIODIVERSITY

ONE-TENTH OF THE earth's surface is covered in ice—about 15 million square kilometers. Collectively, this patchwork of frozen landscapes is known as the cryosphere. It is a massive reservoir of fresh water: ice sheets and glaciers contain 70 percent of the planet's entire supply. Some of these are relatively young, like Alaskan glaciers only a thousand years old, and some are ancient—like Antarctic glaciers up to 20 million years old.

"When we think about the cryosphere, we think about the great ice sheets—the Greenland and Antarctic ice sheets, which are the only two on the planet now—and about mountain glaciers, seasonal snow, and in some cases frozen lake and river ice," says Brian Menounos, a glaciologist and one of the heads of the Hakai Cryosphere Node at the University of Northern British Columbia. "The cryosphere also includes the permanently frozen soil known as permafrost, and Arctic sea ice."

Menounos's collaborator in the Hakai Cryosphere Node is Bill Floyd, an adjunct professor in geography at Vancouver Island University and a provincial research hydrologist. Together they work with the Hakai Institute in researching contemporary glaciers and attempting to understand the changes glaciers are undergoing now, and are likely to undergo in the future. It's a dynamic field, because planetary warming—the result of past and present industrial emissions of greenhouse gases such as carbon dioxide and methane—is causing massive shifts in the cryosphere.

Glaciers, technically speaking, are slowly moving masses of ice that last more than one year. The snow on your lawn is not a glacier, but if it didn't melt and instead grew deeper and larger with each annual snowfall, eventually it would be. Known as "rivers of ice," glaciers exist, at least for now, on all continents except Australia. They are formed in mountainous areas or polar regions by snow accumulating over centuries and millennia. Just because a region remains frozen year-round doesn't mean it will have ice cover. Without snow, you don't have ice. Much of Siberia is cold enough to host long-lasting ice across vast areas, but the climate is too arid to produce enough snow.

When snowfall exceeds melt, the snowpack deepens. It compresses into granules called firn, then into ice. In mountainous areas, eventually gravity and the tremendous weight of the glacier cause the ice to act like a viscous fluid rather than a solid—a phenomenon known as plasticity—and to flow downhill.

The bottom of the glacier grinds along the rocky bed below it, sometimes aided by a layer of meltwater. The glacier also moves internally, as ice crystals slip against each other in response to pressure—an unusual process known as deformation. Because of deformation, not all the ice in a glacier moves at the same rate: the ice in the center moves faster than the ice along the sides. The less dense snow on top lacks the plasticity of the ice deeper down, so it cracks and buckles with the movement, creating glacial crevasses.

Water of life

Some of global warming's consequences are well known, such as ocean acidification (see page 124) and sea-level rise—the latter of which, at current rates, threatens to flood coastal areas and islands in coming decades, from the Maldives to New York City to Bangladesh. But the effects of warming aren't limited to sea level; they range all the way up to the ice fields that cover some of the highest mountains on the planet. Denali in Alaska, Rainier in Washington, Kilimanjaro in Tanzania—all are experiencing glacial retreat.

TOP: Vancouver Island University and Government of British Columbia researchers performing late-winter maintenance and a snow survey at the Homathko weather station southeast of Mount Waddington.

BOTTOM: Meltwater cascading over rocks recently exposed by retreating ice, below the terminus of a small alpine glacier at the head of Bute Inlet.

As researchers watch glaciers shrink at unprecedented rates, warming temperatures are altering the seasonal timing of melting snowpack. This, says Menounos, has serious consequences for aquatic ecosystems like rivers and streams. "Mountain glaciers provide an important buffer for many of the region's headwater streams, both in terms of temperature and volume," he explains. Picture a small creek that's fed by a mountain glacier: that creek will run with lots of cool water during the snowmelt season. "But there's a point at which the seasonal snow has been depleted. And the reason that many of those creeks don't dry up is because they're fed by glaciers. As you melt the snow off the top of the glacier, now you start to melt the ice."

As glaciers disappear, so does the volume of water available to streams in times of drought, along with what Menounos calls "thermal buffering": the protection against heat that glacial meltwater provides. "In some areas, streams will be more susceptible to thermal stress and also more susceptible to drought-like conditions," says Menounos. "Warmer streams and rivers with less water are terrible for salmon, among many other organisms, and may be potentially catastrophic at some level."

On the flip side, river habitats that are currently too cold or icy for juvenile salmon for all or part of the year may see increased salmon populations as temperatures warm and glaciers retreat. Studies indicate that up to 15 percent of salmon don't return to their natal streams, and instead diverge into new spawning habitat. These are the potential pioneers that could colonize rivers newly opened up from ice cover—or whose temperatures have otherwise become hospitable to salmon, rather than prohibitively cold. Whether new spawning habitat opened up by retreating glaciers will be at the same scale as the streams and rivers made uninhabitable to salmon and other fish by global warming is, as yet, unknown.

To better understand what is happening to the cryosphere now—and what may happen in the future—project researchers are tracking glacial snow accumulation, melting rates, and other data. Their measurements and ice-core sampling take them to some spectacular places. Many, like Combatant Col at the base of Mount Waddington and the Klinaklini Glacier that flows from Silverthrone Mountain, are in the Coast Mountains that range from southern

Warming temperatures mean additional stress for many marine and aquatic species, like these juvenile salmon just beginning their journey to the sea.

British Columbia to Alaska, but some are farther east in the interior. British Columbia is an ideal location for glacier enthusiasts: the Coast Mountains alone contain over half of all the glacial ice in western North America, excluding Alaska. As of 2000, that equaled about 2,034 cubic kilometers of ice, which translates to about 825,000 icy replicas of the Great Pyramid of Giza.

Getting onto glaciers or even near them is expensive, tricky work and extremely weather-dependent. When conditions allow, the Hakai Cryosphere Node team conducts ice-core sampling and direct measurements on ice fields and mountaintops. They also use measurements from satellite and airplane LIDAR (light detection and ranging) equipment, which uses laser pulses to measure altitude.

The Airborne Coastal Observatory is a twin-engine Piper Navajo airplane outfitted with high-resolution cameras, LIDAR, and a hyperspectral scanner. Jointly developed by the Hakai Institute, the University of Northern British Columbia, Vancouver Island University, and Kîsik Aerial Survey, the aircraft has enabled researchers to track changes in the size of British Columbia's mountain glaciers.

"We are using airborne laser altimetry, a type of LIDAR, to quantify the changes in the surface elevation of large glacial ice fields," says Menounos. "When we can look at the elevation change with LIDAR and multiply that volume by density, we can figure out what's called mass change of the glaciers—or how much ice or snow is lost or gained. That allows us to quantify how much water is being lost from these frozen reservoirs."

Airborne LIDAR also allows the team to measure how much seasonal snow falls in the Coast Mountains of British Columbia. Pilots fly the terrain prior to the first seasonal snowfall, then take repeated measurements throughout the winter. "This way," Menounos explains, "we can figure out how much snow is being deposited at a particular point in the landscape through time."

Among other things, the plane's hyperspectral scanner can measure the reflectivity of glacial ice. If this sounds like a purely academic exercise, it's not; the ability of a glacier to reflect light has everything to do with its longevity. When glacier algae—which can be pink, red, or green—grow on the surface of ice fields, they speed up melting rates. So does the

The Airborne Coastal Observatory uses high-resolution cameras, LIDAR, and a hyperspectral scanner to observe changes in glacial mass and reflectivity.

TOP: A grand view: researchers take a lunch break overlooking the river-like sweep of Klinaklini Glacier.

BOTTOM: Flying over the Franklin Glacier to inspect snowmelt and ice retreat with a member of the Da'naxda'xw/ Awaetlala First Nation's Guardian Program.

The terminus of the Klinaklini Glacier, approximately 35 kilometers from where the Klinaklini River flows into Knight Inlet. The dark gray appearance of the floating ice is from rock and debris deposited within and on the main glacier from valley glaciers that flow into it.

increasingly common presence of black carbon on ice fields, much of it from forest or bushfires. This is a topic of emerging concern from New Zealand to British Columbia as wildfire seasons have grown longer and more intense in recent years.

"One of the sources of melt in glaciers is the absorption of sunlight," says Menounos. "If it's highly reflective, then it reflects a lot of incoming sunlight. But with the hyperspectral scanner, we can actually look at the changes that can occur following a forest fire. We can see if forest fires are actually causing changes in the reflectivity of ice fields."

Rocks, coffee, and fish

As mountain glaciers melt in western North America, the fresh water they contain swells streams and rivers, and ultimately ends up in the coastal waters of the Pacific Ocean. This water is not just water: thanks in part to glaciers, coastal rivers carry materials that have profound effects on the ocean. As these ice rivers flow down out of the mountains, they behave like sandpaper, scraping and scratching the rocky beds below them, a process known as abrasion. These rivers of ice can also occasionally freeze to knobs of bedrock and pluck away large chunks, which is called quarrying.

Both processes produce abundant sediment, says Menounos. Rock flour from granite and quartz-diorite gives glacierized waters—from the Joffre Lakes in southern British Columbia to the Klinaklini River, which empties into Knight Inlet on the province's Central Coast—a cloudy, turbid appearance. In some waters it imparts a striking turquoise-blue color, but rock flour doesn't just affect aesthetics. These sediments build unique river delta ecosystems and stay suspended in fjord waters, where they block light and limit the growth of phytoplankton. The minerals in rock flour are also important for coastal ecosystems.

"Granites or quartzites can contain a certain amount of phosphorus, which is a fertilizer," says Menounos. Phosphorus and iron liberated from rock are required nutrients for marine food webs, starting with the microbial invertebrates at the bottom (see page 39). "So these glaciers produce not only sediments, but bioavailable nutrients that allow the organisms living in these marine and aquatic ecosystems to flourish."

Carbon is well known as the building block of life for its unique capacity to bond with elements like nitrogen, hydrogen, and oxygen and form stable molecules. In contrast to the turbid waters that flow into the ocean from glacial meltwater, nearby rain-fed streams are nearly free of sediments, yet carry highly concentrated carbon from forests and wetlands.

The disappearance of glaciers is expected to have a variety of consequences for species that depend on rivers and streams, like these chinook salmon.

"I think of it a little bit like a coffee maker," says Ian Giesbrecht, an ecosystem scientist who coordinates the Hakai Institute's watershed science program. "You pour nice, clear-looking water into the machine and it percolates down through the ground-up organic material of those coffee beans. Out the bottom comes this dark, rich water that's got a high concentration of dissolved organic carbon."

In a similar way, Giesbrecht explains, in low-lying terrain with wetlands and organic soils, rain percolates down through the ground, steeping in a rich organic environment of branches, leaves, moss, and soil. "That's the organic material that's been sitting in the soil for years and years and then makes its way downstream carrying a lot of carbon," he says. "It can have a brownish or slightly orange-brown tint to it, which is coming from the organic matter dissolved in the water."

Glacial rivers deliver important mineral nutrients and organic matter that feed freshwater and marine food webs.
Rivers in forested areas with wetlands contribute more concentrated organic matter to these systems.

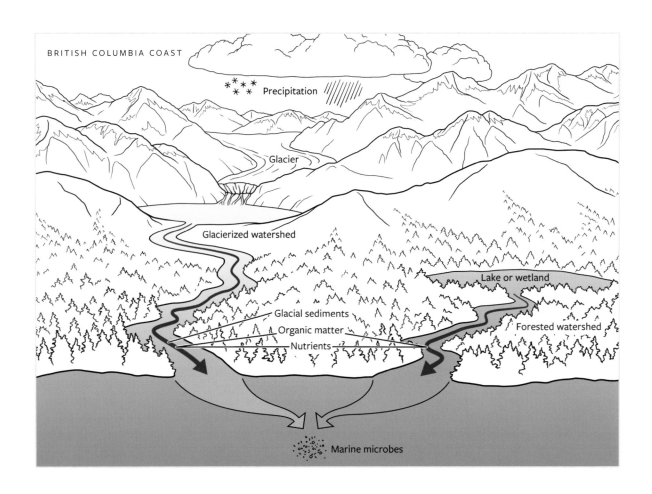

That dissolved organic matter, known to researchers as DOM, is a food source for organisms in marine and freshwater ecosystems. "We see evidence of this dissolved organic matter getting into the estuaries on Calvert Island," says Giesbrecht. He cites the work of researchers Colleen Kellogg and Kyra St. Pierre, noting that their studies show how the presence of DOM in these ecosystems "stimulates production of the microbial community and changes the composition of that community."

Glacial meltwater also carries carbon into the ocean. While the concentrations of carbon in meltwater are lower than in waters that have percolated through forests and wetlands, the huge volume of it makes a significant contribution to coastal oceans in the northeast Pacific. This includes ancient organic carbon—mainly from microbes living on and under the ice. "It's been shown elsewhere that terrestrial carbon from ice fields can be traced right through to the marine food web and even into the bodies of birds that are living in Glacier Bay in Alaska," says Giesbrecht.

View from the terminus of a small glacier, looking down Bute Inlet. Bedrock newly exposed by melting is visible just below the retreating ice.

Reflecting on change

At the Klinaklini weather station, researcher Trevor Dickinson of Vancouver Island University's Coastal Hydrology and Climate Change Research Lab conducts a drone survey to create a 3D model of the surrounding area.

In the heavily glacierized coastal regions of British Columbia, rivers and streams swell and warm in the spring and summer as seasonal snow and ice melt. In areas of the coast distant from snowmelt and glacial runoff—as in the islands of the outer coast—water levels peak with the heavy rainstorms of fall and winter; streams are at their lowest in the warmer months. Like Menounos, Giesbrecht notes that as glaciers are melting and threatening to disappear altogether, this traditional timing is changing. "Climate change is playing out differently in these systems. The expectation is that we're going to see a shift from timing driven by melting ice and snow toward more rain-driven timing."

More rapid ice-melting may lead at first to higher and colder river flows in the summer. As glacial ice continues to shrink, less meltwater and warmer temperatures will be the likely results in many rivers, with consequences for salmon and other aquatic species. There may be winners and losers: new aquatic habitat potentially opening up in northern areas, while other rivers and streams become too warm for some of their native species. Human use of river water for hydroelectric power and other industrial purposes will also be affected by glacial retreat. And there are other dimensions that are even more difficult to prepare for, or to quantify.

"It's not really scientific, but one of the aspects of my work with glaciers is a chance to appreciate how remarkably beautiful ice is, and how unfortunate it is that it now seems it will be in our mountains for a limited amount of time," says Menounos. "Certainly I'll be long gone before the ice is gone, but my kids' grandchildren will likely see the Coast Mountains without ice. Considering it's taken five or six thousand years of Earth's history to form that ice, to think of it disappearing in a hundred years is rather shocking."

View up the valley of a tributary flowing into the Klinaklini River, visible in the foreground.

3 SMALL IS BEAUTIFUL
MARINE FOOD WEBS

IF YOU'RE AN Adams River sockeye salmon (*Oncorhynchus nerka*), by the time you reach Discovery Passage you've come a long way. Hailing from spawning grounds far inland in British Columbia, sockeye from the Adams travel over 400 kilometers downstream to get to the sea. From there, the mouth of the Fraser River, it's another 180 kilometers to get to the entrance of Discovery Passage.

The salmon smolts are either one or two years old when they make the trip. At around 12 centimeters long, they're a little shorter than a ballpoint pen. Having made it to Discovery Passage—which, proportionally, is like a human swimming from Seattle to Peru—they still have a long journey ahead of them. The Gulf of Alaska, where they'll grow to maturity, is another thousand kilometers away. After two to three years feeding and fattening themselves in the open ocean there, the tolling of an inner bell will send them heading back on the long return trip.

Surviving to spawn and begin the cycle again, of course, requires that they avoid getting eaten along the way. Predators happy to make a meal of sockeye at various points of their life cycle include double-crested cormorants (*Phalacrocorax auritus*), harbor seals (*Phoca vitulina*), salmon sharks (*Lamna ditropis*), and the abundant and highly successful species known as humans (*Homo sapiens*). There are many other animals that prey on sockeye—and along with these, some lesser-known but important factors in their survival.

Salmon are part of the marine food web, eating organisms and being eaten in turn. For sockeye to be alert, fast, and strong enough to evade being a meal for another creature before spawning—and to make it all the way back upstream to the Adams River—they have to be well fed.

So what do salmon eat? And what other, less obvious food-web factors affect salmon in the ocean? Questions like these are a submarine port-hole onto a world few people see. It's not at the scale of sharks and seine nets; it's much smaller, ranging from copepods a few millimeters long to viruses so small they can be seen only with an electron microscope. This understory of the food web is a complex and beastly world, rich with life and kept in check by an abundance of fearsome predators. Researchers are just starting to understand it, and some of its qualities are startling.

Going deep at QU39

Discovery Passage is at the northern tip of the Strait of Georgia. Not far away, just off the southern tip of Quadra Island, is a point known to researchers as QU39. You could row right over QU39 in a wooden dinghy and not notice, because there's nothing there: it's a GPS point, calculated to the 10,000ths of a degree for both latitude and longitude. The waters that surround it are of interest to science because the area is a bottleneck for migrating sockeye, among other salmon species.

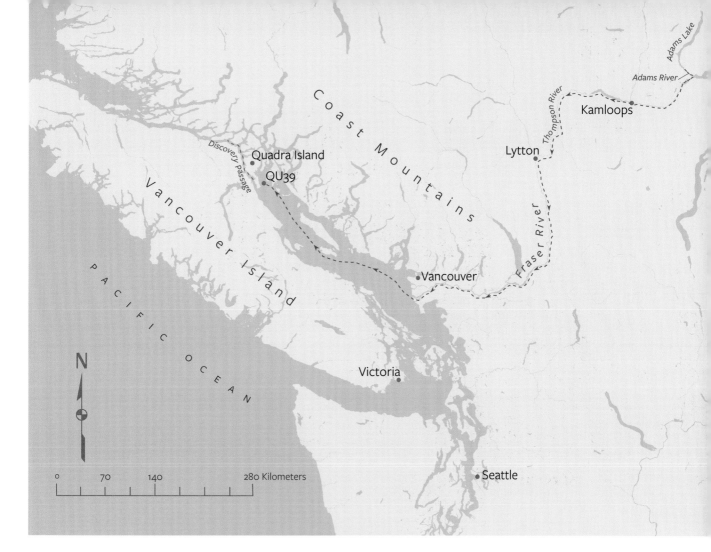

Coast Mountains

Adams Lake

Adams River

Kamloops

Thompson River

Lytton

Discovery Passage

Quadra Island

QU39

Vancouver Island

Fraser River

Vancouver

PACIFIC OCEAN

N

Victoria

Seattle

0   70   140   280 Kilometers

Adams River sockeye travel great distances out to sea and back again. The trip from their natal streams to the Quadra Island waters monitored by Hakai researchers is about 580 kilometers.

"Where we work in the northern Strait of Georgia is a really important area for juvenile salmon," says Colleen Kellogg, an oceanographer and microbial ecologist at the Hakai Institute. "These salmon feed mostly on zooplankton, and herring, which are important to salmon, eat zooplankton as well. But if phytoplankton weren't growing in the area, and if bacteria weren't recycling nutrients, then the zooplankton wouldn't have the food that they need. And if the zooplankton don't have the food that they need, then they wouldn't grow to the numbers that the salmon need to eat and survive."

What this means, in short, is that salmon are part of a far more complicated food web than it may appear—and their fate depends on a teeming population of things that are mostly invisible without a good microscope.

A food web is made up of the interactions between organisms in an environment. Plants or plant-like organisms create food from carbon and sunlight (or chemical reactions) and other organisms feed on them—and on each other. The result in any ecosystem is a complex web of nutrient flows and predator-prey relationships. This can look like anarchy or order, depending on your point of view. Kellogg points out that food webs are everywhere, whether deep in the soil, in the ocean, or on land.

"We're all part of a food web," she says, noting she and her colleagues are attempting to better understand the particular relationships that underpin the food web near QU39. "We're looking at organisms from bacteria through to zooplankton, and developing methods to look at what the zooplankton might be eating."

If organisms like bacteria and zooplankton seem unimportant because they're tiny, biological oceanographers will assure you that they're not. Current research indicates that in the ocean, over 90 percent of the living material by weight—known as biomass—is microbial. Phytoplankton, zooplankton, and bacterioplankton are some, but not all, of the miniscule life-forms that make up this deep layer of the marine food web.

The goal of QU39 researchers is not just to identify what they find in the samples they take there, but to begin to understand what these planktonic organisms do with, and to, each other. Brian Hunt is an ecosystem oceanographer (see page 113) who works with Kellogg on food-web research at QU39. He cites "a common statistic" that in a drop of seawater—about one milliliter—you can find 1,000 protists (protozoans and algae), 1 million bacteria, and 10 million viruses. That is a lot of microbiology underpinning the health of salmon and other marine creatures, and a lot of interactions. Despite the abundance, observing how such

FACING: Top, the health and survival of juvenile salmon like these are inseparable from a complex microbial food web that includes zooplankton, bacteria, protists, phytoplankton, and viruses. Bottom, the biomass of returning salmon provides a wealth of nutrients for aquatic and terrestrial creatures in food webs beyond the marine environment.

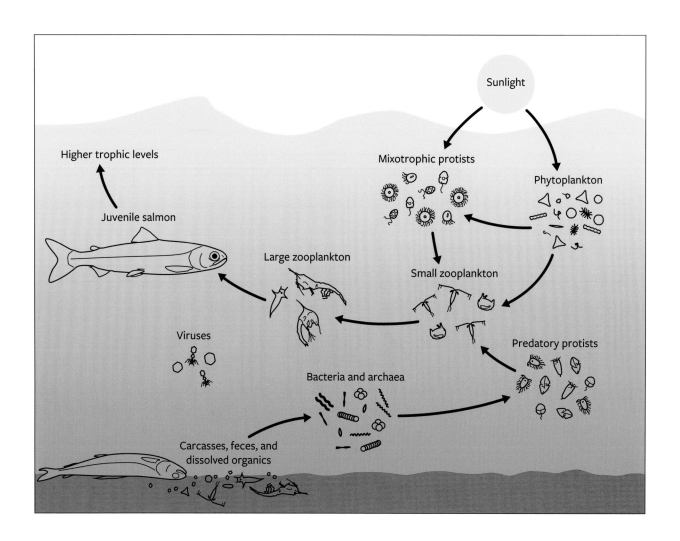

Sunlight

Mixotrophic protists

Phytoplankton

Higher trophic levels

Juvenile salmon

Large zooplankton

Small zooplankton

Viruses

Predatory protists

Bacteria and archaea

Carcasses, feces, and
dissolved organics

This simplified version of the marine food web in the Salish Sea shows some of the interactions that make life possible in the ocean.

organisms interact is difficult even with the most advanced microscopes. Generally, says Hunt, when researchers observe microbial plankton, everything they're looking at in their samples is already dead. And while DNA testing is invaluable in their work for tracking what is present in the water, it doesn't do much to help researchers understand behavior, except in a few specific applications.

A "first-order" way of identifying relationships and working around this, Hunt says, is to get frequent samples and track abundance levels. If the population of a certain type of algae plummets every time that of another planktonic species spikes—whether virus, bacteria, protist, or zooplankton—you can surmise that the burgeoning species is killing off

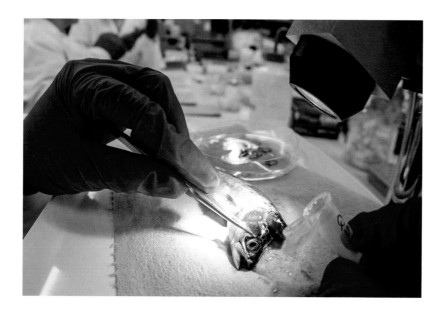

the algae. This gives you a hint at one not-so-subtle relationship that you can then investigate further. "Once you've identified a correlation, where you see something always occurring together, or where there are opposite cycles, we can put them in an experimental system and test that relationship directly," says Hunt.

Beyond testing correlations, some work with microscopes can be helpful, particularly for things that have hard body parts, like diatoms—algae that live in transparent silica shells. It's even possible, says Hunt, to use genomic testing to identify what larger zooplankton have been eating by sequencing the DNA of the contents of their gut.

Worlds within worlds

One thing to understand in all of this is that the term "plankton" doesn't refer to a species or to something only baleen whales eat. It is actually a vast grouping for any and all marine organisms that don't have self-propulsion or are too small or weak to resist the flow of ocean currents and tides. If you're a type of plankton, you are, literally, a drifter. Eggs, larvae, or immature organisms that are carried on the currents are included. So are jellyfish. Even the lion's mane jellyfish (*Cyanea capillata*), whose 30-meter tentacles make it longer than a blue whale, is considered a type of plankton.

LEFT: Researchers dissecting a juvenile salmon for analysis.

RIGHT: Marine plankton are not only foundational to ocean food webs, but also fascinating to look at when magnified. Top, an *Asteromphalus* diatom. Center, an *Emiliania* coccolithophore. Bottom, a *Dinophysis* dinoflagellate.

Not all plankton are tiny. Lion's mane jellyfish are technically a type of plankton, though their tentacles can stretch to the length of a blue whale.

Phytoplankton are plant-like ocean drifters. This means (in most cases) they are autotrophs: organisms that produce their own food through photosynthesis or chemical energy. Zooplankton, by contrast, are animal-like organisms, known to science as heterotrophs. To survive, they need to consume phytoplankton or other zooplankton, or the fine, slowly sinking detritus of dead organic material known as marine snow, which is everywhere in the ocean. An example of a type of zooplankton that feasts on marine snow is copepods, tiny shrimp-like crustaceans that are a favorite snack of salmon.

Bacterioplankton as a category encompasses single-celled bacteria as well as archaea—a type of single-celled organism once associated mainly with extreme environments like hydrothermal vents, but now known to be abundant throughout the water column. Viruses, which are hard to pin down at the best of times and whose existence challenges basic definitions of life—scientists are still split as to whether they are alive or not—are in a planktonic category of their own. Together, they are consequential for larger ocean fauna like salmon.

"There's a huge number of species and different forms of life that are interacting, from bacteria, viruses, and protists, through to the zooplankton," says Hunt. "The end product is your krill, or your big fat copepod.

And that's a juicy thing that the salmon is munching on."

New discoveries about microbial behavior have blurred the categories that divide microbial plankton, however. The trouble with the phyto-zoo-bacterio breakdown is that we now know that many plant-like phytoplankton eat other organisms—a behavior known as mixotrophy—and thus span the phytoplankton/zooplankton divide, and that many phytoplankton are actually a kind of photosynthetic bacteria called cyanobacteria (which have been around in some form for about 3.5 billion years).

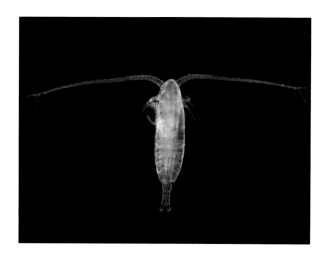

Calanus pacificus copepods like this one are a favorite snack for Pacific salmon.

One consistent way to bring some order to planktonic organisms is to categorize them by size, from the extremely minute to the very large. The smallest plankton are measured in millionths of a meter, known as micrometers or microns. (To give some idea of what this means, a human hair is about 60 microns thick.) Femtoplankton, made up entirely of viruses, are the smallest of the bunch, measuring less than one-fifth of a micron. It would take 300 of the largest femtoplankton-class viruses to span a human hair. Some viruses are larger than this, up to a full micron, and classed in a category called picoplankton, but they're still so small that in most cases scientists need an electron microscope to spot them. At the other end of the plankton-size spectrum are the megaplankton, drifting organisms over 10 centimeters long, such as many jellyfish.

Bigger things tend to eat smaller things, so the size of an organism can tell you something about its feeding habits. While size doesn't tell you much else that is meaningful about behavior, it does offer a way to sort and compare. And a close look at this layer of the food web through the lens of size offers some surprises.

"The size range of phytoplankton expands along five orders of magnitude," says Hunt. "The size range in a plankton community is about three times the size range between a herring and a whale. It's just enormous. Because they're all small, we aren't able to see or appreciate the nuance, but if you zoom in with a microscope, there are worlds within worlds of things that are going on."

As might be expected, a lot of what is going on in this world is things digesting, eating, or infecting other things. Bacteria and phytoplankton

are foundational to the system. All life, and all food, is carbon-based, and
these organisms are responsible for taking up organic and inorganic carbon (see page 33) into their cellular structure. As they do, they make it
available to other creatures in the food web. Marine bacteria, Kellogg
points out, also break down organic matter and make various compounds,
including vitamins, available to organisms like phytoplankton. "Without
bacteria we wouldn't have these productive phytoplankton blooms that
are essentially the base of marine food webs as most people think of
them."

Bacteria, and archaea, are eaten by protists and smaller zooplankton. Zooplankton feed on phytoplankton. Some bacteria feed on other
bacteria. Viruses attack bacteria, phytoplankton and other protists, and
zooplankton. "That all may seem really negative," says Kellogg with a
laugh. "Mostly a lot of what's going on in the ocean is very happy, and
there's a lot of cooperation among all of these organisms."

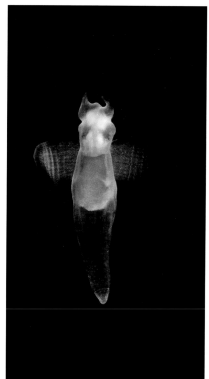

TOP: Megan Foss showing a plankton sample.

BOTTOM: Colleen Kellogg performing genomic testing at the Marna Lab.

Modeling the future

Hakai Institute researchers have been heading out weekly to QU39 since late 2014. To capture seawater for analysis, a team of three to five people sends down from the boat a long string of oceanographic sampling devices called Niskin bottles, which descend to a maximum depth near the seafloor of about 250 meters. This is fairly deep, more than two and a half times the height of the Statue of Liberty. The end of the line is equipped with a sensor called a CTD, which measures, among other things, the conductivity (a way to gauge salinity), temperature, and depth of the seawater as it descends, six times every second. When the bottles are brought back, water is collected for DNA sampling and other tests, including measuring the nutrient and carbon concentrations. Kellogg and her team extract the environmental DNA, or eDNA, out of each QU39 sample and then compare the genetic material to a reference database, allowing them to "see" what is present in the water at that moment.

Relative to other food-web research efforts around the world, says Kellogg, this weekly sampling to a wide range of depths is providing researchers with an unusually robust amount of data.

"What I'm seeing is that there are times in the year when microbial populations are quite stable, as with bacteria and archaea in the winter," says Kellogg. "Then once spring kind of turns on in the ocean in late February or early March, that 'who's who' in the water can change from week to week."

When you're studying and taking samples on a monthly timescale, she adds, "you can miss the changes in communities that you can spot through this finer resolution."

Kellogg hopes that data on the physical and biological factors will allow her to build environmental models and forecast possible changes to the marine food web due to climate change–related phenomena such as marine heat waves (see page 120).

"By sampling so frequently now, well, it means that if you can posit what the temperature and salinity are going to be in these waters in twenty years, I might also be able to predict what type of bacteria we'll have there," she says. "Hopefully by then we'll also know the main function of those bacteria in the system—and then we can know how changes to temperature, to pH, and to other things will affect the ocean food web as time goes on."

All about the fat

Animals depend on the food that they eat to survive, and what's available as food is impacted by environmental conditions. The chemicals and nutrients that ocean life depends on are affected by temperature, by fresh water that comes into the ocean, and by pH (the acid or alkaline quality of water), among other factors. These in turn can be affected by seasonal factors and multiyear processes like El Niño and marine heat waves. Creatures in the food web are also affected by events that might begin at the microbial level but have macro-scale effects—like sea star wasting disease or other ocean epidemics (see Chapter 4).

Brian Hunt sampling plankton with a bongo net off Quadra Island.

LEFT: The *Hakai Spirit* research vessel in the waters off Quadra Island.

RIGHT: Bryn Fedje and the Hakai oceanography team sampling plankton.

But just because something is there to be eaten doesn't mean it's always a great meal. In other words, says Hunt, not all copepods are created equal. "We talk about prey quality quite a lot," says Hunt, explaining that one of the critical components of prey quality is fat content. "We care about what we eat, and that applies to everyone, even the little guys in the ocean."

Fatty acids like omega-3s are important for humans, but also for other animals, including fish. Young salmon need a diet rich in fatty acids, particularly two known as DHA and EPA. Without them, salmon's neurological system won't develop properly, and they won't have the acute vision, coordination, and response mechanisms necessary to make the epic trip from spawning grounds to open ocean and back again. "Everyone's worried about the survival of juvenile salmon, and people often talk about seal or bird predation," says Hunt. "But we can't forget the really strong connection between what they eat and their health."

This same reasoning, says Hunt, is why viral diseases are a significant factor in food webs. In the case of salmon, while a viral disease itself might not kill them outright, it can weaken them and impair their performance, making them less able to escape predators.

QU39 researchers measure the health of the juvenile salmon that are coming through the area, and explore links between the smolts' well-being and the larger food web—as well as physical ocean conditions like temperature. Not long after the QU39 research began, the marine heat wave known as the Blob hit British Columbia's coastal ocean (see Chapter 7). Warm waters that lasted for nearly three years "really impacted" the salmon, says Hunt. One clear link can be summed up by saying that

colder waters are, in general, good for salmon: when ocean temperatures go up, the fatty-acid content of the copepods and other zooplankton on which salmon feed goes down. Less nutritious prey is bad news. During the marine heat wave and its aftermath, young smolts at QU39 were unusually skinny and in poor health.

"Ocean conditions affect the nutrients available to all the creatures in it," says Hunt. "That's why in our work we go really right down into the weeds, right to the beginning where the physics and the chemistry of the ocean are creating a substrate for life. This is the black box, the productivity-generating machine that makes it possible for things we care about, like salmon, to eat and survive."

But the machine can't be separated from the organisms that depend on it, and impacts go both ways. While the microbial food web is directly influenced by ocean conditions, planktonic microorganisms influence the ocean in turn through processes such as ocean acidification (see page 124) and the carbon cycle—the way organic and inorganic carbon are exchanged between the atmosphere and the ocean and taken up by organisms. This is another aspect of the microbial world that QU39 researchers are exploring.

Brett Johnson sampling juvenile sockeye salmon in Okisollo Channel.

GOING VIRAL
A Q&A WITH CURTIS SUTTLE

Curtis Suttle is a biological oceanographer and an expert in marine virology. A professor at the University of British Columbia, his research focuses mainly on viruses and their role in the environment. He has examined viruses in the oceans, in the high Arctic, in deep mines, in lakes, and even 3,000 meters above sea level in Spain's Sierra Nevada mountains.

Q: Before we get into talking about ocean viruses specifically, can you tell me a little about the marine microbial world that they're part of?

A: Food webs are complex. There are many linkages in them, and viruses are critical to the whole system. If you think about oceans and their food webs, the first thing to know is that they're microbial. Most people don't realize that 90 to 95 percent of the living material in the ocean by weight is microbial.

Viruses are the most abundant biological entities in the oceans; in the microbial world, they outnumber marine bacteria by about 10 to 1. That adds up to a big number. In a liter of coastal seawater there are typically about 10 billion viruses. If you take the minimum number usually found and multiply that by ocean volume, you get about 10^{30}, a number called a nonillion. If you put all those viruses end to end, they would stretch millions of light-years into space, farther than the nearest 60 galaxies.

Q: That's staggering. I had no idea. It seems obvious then that viruses play a pretty important role in the ocean.

A: First of all, it's important to understand that viruses can't replicate on their own. They have to actually infect something in order to propagate, and they are very host-specific. One type of virus tends to only infect a very narrow range of organisms—and usually only particular genetic strains within a species. Predation or grazing by other organisms in the food web tends to be more general, attacking or feeding on a wide variety of things. Viruses tend to be highly focused and specific.

Viruses are critical in the ocean because mortality is the key to food webs. In order for life to exist, there has to be death—where organisms become food for other organisms and populations are kept in balance. In ocean environments, viruses are key agents of mortality whether you're looking at bacteria, fish, or marine mammals. They are responsible for killing about 20 to 30 percent of the living material—again, by weight—in the oceans every day. Most of that is bacteria.

Q: That's another shocking number. Twenty to 30 percent of the ocean's biomass dies every day?

A: Yes, and one of the ways that happens is that viruses control blooms of microorganisms. The dynamics of viral infection mean that the more dense the host population is, the faster viruses are able to propagate. For example, one of the big formers of blooms that we have in coastal Pacific regions is a type of algae, *Heterosigma akashiwo*, which is a fish killer.

There are several different viruses that will rise and kill *Heterosigma akashiwo*. When populations get really dense, viruses will propagate quickly and in many cases cause that population to collapse—which is why a bloom situation is actually pretty unusual, because viruses are working in the background all the time. When a certain organism or phytoplankton gets really dense, that's when viruses are likely to become most important. Viral infection occurs very quickly and very rapidly, and so can decimate a population in some cases in hours or days.

When a virus kills a bacterium, protist, or other organism, it releases or liberates those nutrients from that host organism back into the water, something called the viral shunt. This makes them major players in the biogeochemical cycling in the ocean and in supplying nutrients to other organisms. In fact, we've done experiments where we've taken viruses out of the system and in some cases photosynthesis stops. So again, in order to have life, you have to have death, right? It's as simple as that.

Q: It's a rather different way of looking at viruses, seeing them as essential to life. Humans tend to regard them only as something to defend themselves against, as in the COVID-19 pandemic and other such outbreaks.

A: Well, of course, we can't really keep viruses totally at bay. Every time we inhale, we're breathing in probably tens of thousands of viruses. That's back to the point that viruses are so specific in targeting what they infect. The vast majority don't make us sick. In fact, a wonderful paper came out a few years ago suggesting that the viruses we inhale are actually a very important part of our immune system, because they coat our lungs and attack pathogenic bacteria that we inhale and that would make us ill if viruses didn't kill them.

Humans would not exist if it weren't for viruses. We're absolutely dependent on them in terms of their role in the ecosystem, but we're also intertwined with them in our evolutionary development. Eight to 10 percent of our entire genome is actually viral, derived from viruses trapped in our own genome. As one example, an important protein in the placentas of mammals is encoded by a gene that was donated by a virus to our ancient ancestors. There would be no placenta in mammals if it weren't for viruses. So we can't ever really separate our existence from theirs. ∎

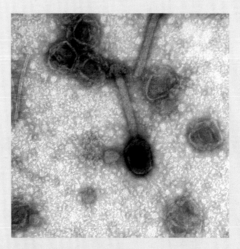

Electron microscopy image of a cyanophage, a virus that feeds on cyanobacteria, a type of phytoplankton found in marine and fresh waters.

4 SEA STARS

A KEYSTONE SPECIES AND MARINE DISEASE

IN 1963, A young researcher named Robert "Bob" Paine turned the study of ecology on its head by using a tool not often associated with such research: a crowbar.

Paine was 30 years old, a recent hire at the University of Washington as an assistant professor of zoology. His innovative approach consisted of prying orange and purple sea stars (*Pisaster ochraceus*, also called ochre stars) from shoreline rocks and flinging them back into the chilly waters of Makah Bay. While anyone witnessing it might have thought Paine's behavior bizarre, his experimental approach was methodical and precise— and it would eventually revolutionize the way we think about animal communities, from Pacific kelp forests to the African savannah.

In a departure from the purely observational research that dominated ecology in those days, Paine was trying a "kick it and see" approach. He wanted to see what would happen to biodiversity if he removed a top predator from a particular area. Inspiration for the experiment came from his academic adviser and mentor, Fred Smith, who (with two other colleagues) theorized that carnivorous predators controlled populations of plant-eating animals farther down the food chain. This was new thinking: most ecology researchers believed that predator populations were

PREVIOUS SPREAD: Purple and
orange ochre sea stars, black
katy chitons, and other sea
life off Calvert Island.

controlled from the bottom up, limited by the number of prey available to them.

It may not sound all that controversial, but Smith and his colleagues found their papers on the subject roundly criticized. What was needed to prove the vital role of predators in biodiversity was evidence. Thanks to a stretch of beach on Makah tribal lands in Washington State, Paine was able to provide it.

Paine first encountered Makah Bay on a field trip with University of Washington students. As molecular biologist Sean B. Carroll describes it in a review of Paine's work, the spot was a rich playground for biology research:

> The tide pools were full of colorful creatures—green anemones, purple sea urchins, pink seaweed, bright red Pacific blood starfish, as well as sponges, limpets, and chitons. Along the rock faces, the low tide exposed bands of small acorn barnacles, and large, stalked goose barnacles, beds of black California mussels, and some very large, purple and orange starfish.

These ochre stars, which can appear in shades of orange, purple, and even brick red, are a top predator in the intertidal zone, the area of shoreline that is exposed only during low tides. In the rocky areas of Makah Bay, they feasted on limpets, snails, barnacles, chitons, and, most critically, California mussels (*Mytilus californianus*). Paine's experimental removals of ochre stars took place on a tiny sliver of the six-kilometer shoreline of Makah Bay: a seven-meter section of rock almost two meters high. Next to it was an area that he left undisturbed as a comparative example called a control site.

From 1963 to 1966, Paine returned to Makah Bay at least once a month, crowbar in hand. He flung an untold number of sea stars into the bay, tallying the statistics of 15 species of intertidal life—from seaweed to chitons—as he did so. Within 12 months, the community of organisms in Paine's study area showed surprising changes. Goose barnacles and mussels had displaced smaller acorn barnacles. Four species of seaweed had disappeared, along with two types of chitons and limpets. Populations of sponges and anemones were down as well.

These were unexpected results that would turn out to have major significance. After only a year, the loss of a critical predator had reduced biodiversity on Paine's test area by 50 percent, and over time, the

impacts became even more pronounced. As Paine observed in years to come—in Makah Bay and in decades of research on nearby Tatoosh Island— California mussels took over the rocks completely, pushing out all the other species he had been tracking. The resulting mussel beds did offer foundational habitat for a diverse array of tiny organisms that lived amidst their shiny black shells; however, the difference between the experimental and control areas was stark. Without ochre stars to keep them in check, mussels formed a monoculture that excluded almost all the species that formerly resided in that area.

Paine would go on to explain his findings more simply with the notion of a "keystone species." A keystone is the central, wedge-shaped rock in an arch; when securely fitted, it locks the other arch stones in place. "If you remove the keystone of an arch, the whole arch collapses," Paine told *Hakai Magazine* in 2015. "That's exactly what these natural systems do. And so I figured, 'Well, why not just call it a keystone?' And people picked up on it."

LEFT: Mussel predation by an ochre sea star.

RIGHT: Ochre sea stars (*Pisaster ochraceus*) ringing a tide pool.

Bob Paine exploring the intertidal zone at Makah Bay, Washington.

In any given area, there are species that have a disproportionate effect on the overall diversity of animals and plants. When they're removed or suffer a die-off, diversity can collapse in what is called a trophic cascade. Extirpations—local extinctions—of species are a frequent result.

Since the 1960s, researchers have used Paine's pioneering work to identify keystone species in habitats all around the world. At Makah Bay and Tatoosh Island, that keystone was the ochre star. In the Greater Yellowstone Ecosystem, which stretches across Montana, Idaho, and Wyoming, gray wolves (*Canis lupus*) are a keystone species that affect the abundance of a wide variety of organisms, from sedges to songbirds. In areas of the Pacific coast where it has rebounded from near-extinction, the sea otter (*Enhydra lutris*) is a keystone species critical to the creation and health of kelp forests (see Chapter 5).

In recent years, ochre stars and other west coast sea stars have again become a focus of scientific research—this time as part of an inquiry into a mysterious disease that spread like wildfire throughout the waters of the northeast Pacific.

Rise of an epidemic

In the autumn of 2013, marine ecology researchers began to hear troubling reports from BC scuba divers. On the seafloor of Howe Sound and other inlets in southern British Columbia, sea stars—popularly known as starfish—were beginning to literally melt away. Divers saw species like the sunflower star (*Pycnopodia helianthoides*) twisting and contorting themselves into gruesome shapes before developing lesions and falling apart—partly through autotomy, a desperate predator-evasion measure of detaching their arms. The infected sea stars then essentially melted in place, their tissues dissolving and leaving only a ghostly pile of bony ossicles.

Only three years before, divers had noted unusual "swarms" of sunflower stars in some of these same areas. Underwater photographs showed thousands of sunflower stars congregated into a dense, psychedelic carpet of orange and reddish purple. This was an unknown phenomenon, one that researchers were at a loss to explain. Most observations of sunflower stars showed them being far more solitary in their behavior, coming together only in much smaller groups to facilitate mating.

The puzzle of swarming sea stars turned out to be short-lived, however, erased by the newer, opposing mystery of a disease outbreak taking a catastrophic toll on sea stars from Baja California to southern Alaska. In this area known as the northeast Pacific, the disease infected over 20 species; hardest hit were sunflower stars, mottled stars (*Evasterias troschelii*), morning sun stars (*Solaster dawsoni*), giant pink stars (*Pisaster brevispinus*), and the sea stars made famous by Bob Paine: ochre stars.

Christopher Harley, a professor in the University of British Columbia's zoology department, had been studying ochre stars for over a decade by that point. His research included exploring the ecological origins of their color variation. Ochre stars in the inland waters of the Salish Sea are mostly a brilliant purple, but elsewhere on the coast they can vary from pale orange to brick red. Along with published papers, Harley kept regular tallies of the sea star populations he'd observed clinging to rocks in the intertidal zone in the Vancouver area.

A dense aggregation of sunflower stars at Knight Inlet, British Columbia, in 2014.

ABOVE: Left, Christopher Harley collecting specimens for an experiment on the interaction between mussels and endolithic cyanobacteria that erode their shells. Right, Harley taking images of an experiment evaluating competition between two types of dog whelk sea snails (genus *Nucella*).

FACING: Typical tide pool from the outer coast of British Columbia prior to the sea star wasting disease outbreak that began in 2013.

"I'd been keeping tabs on sea stars for a long time because they're very important ecologically, and they're also incredibly easy to count," Harley recalls. "If you're a researcher at a beach around Vancouver, or all the way out on Calvert Island in BC or Tatoosh Island in Washington, and you've got an extra 10 minutes, you may as well count ochre stars because they're a pretty good indicator of what else is happening in the system."

Fresh water can kill sea stars, and in the Vancouver area, salinity levels drop in summer due to meltwater outflow carried by the Fraser River. The result, says Harley, is that late fall and winter are the best times to see sea star populations in intertidal areas around Vancouver. (The opposite is true of most rocky coastlines between Baja California and Alaska, when rough surf from winter storms causes ochre stars to retreat away from the intertidal zones to deeper water.) As evidence of a disease outbreak continued to mount over the late fall and winter, in December of 2013 Harley went to visit some of the rocky Vancouver beaches that previously had shown robust winter populations of ochre stars. Several appeared normal, or close to normal. One was not.

"On my previous survey at Third Beach in Vancouver's Stanley Park, there had been about one thousand ochre stars," Harley says. "On my visit in late 2013, there was only one. There wasn't even any evidence they had been on that beach. They had completely wasted away."

In the ensuing months, formerly ubiquitous ochre stars were decimated at four of the five local beaches Harley frequented. The scene

Sunflower star (*Pycnopodia helianthoides*) succumbing to sea star wasting disease.

was repeated at a site near Nanaimo on Vancouver Island called Yellow Point. "I arrived at Yellow Point right near the peak of the actual outbreak," he recalls. "And I hadn't been prepared for the emotional impact of it. You'd see arms breaking off from the dying bodies and crawling away. I'm a fairly rational, scientific person, but I had spent a lot of time studying these very interesting animals, and watching them suffering this horrible fate was really depressing."

Researchers up and down the coast were witnessing similar events, attributed to a syndrome known as sea star wasting disease (SSWD). It soon became clear that they were observing the unfolding of a massive epizootic—the term for an epidemic in the animal kingdom—one many researchers believe is the largest known marine disease event in recorded history. Between 2013 and 2015, SSWD spread with devastating speed, killing millions of sea stars in over 20 species across a range of several thousand kilometers. Even sea stars in public aquariums weren't safe. All the aquariums on the West Coast that replenished their tanks with unsterilized seawater—a standard and usually reliable practice—watched as their captive sea stars died, many while still on public display.

In a little over two years, formerly abundant sunflower stars were brought close to extinction. Researchers working with the Hakai Institute estimated that sunflower star populations in British Columbia's Central Coast had declined by 96 percent by 2014; by 2017 their numbers in the rocky reef habitats of the region were "virtually zero." In the aftermath, scientists in a variety of disciplines have been working to understand the causes of the epizootic and the effects it is already having on marine ecosystems on this coast.

From the beach to the midnight zone

The class of organisms known as sea stars dates to the Ordovician period about 500 million years ago. Sea stars are also called asteroids after the name of their taxonomic class, Asteroidea. Their relatives in other classes in the phylum Echinodermata—alluding to their spiny skeletons—include, among others, sea cucumbers, urchins, and the more recently discovered deep-dwelling sea daisy.

Worldwide, there are over 1,900 species of asteroids in 36 families. They can have five arms, like the blood star (*Henricia leviuscula*) found off the west coast, or up to 50, like *Labidiaster annulatus*, a krill-feeding sea star from Antarctic waters. They're found on the ocean bottom everywhere from shallow coastal areas to the lightless depths known as the midnight zone. One of the deepest known sea stars, *Eremicaster vicinus*, is a small mud-dwelling asteroid that lives over 7,600 meters—that's 25,000 feet—below the ocean's surface.

Half of the world's known sea star families are represented in the northeast Pacific. Many species, such as sunflower stars and ochre stars mentioned earlier, are endemic to these waters—that is, found nowhere else on the planet. The complex, glacier-scoured landscape of rocky coastal fjords and islands from Washington State's Puget Sound to southern Alaska offers an especially rich habitat for sea stars. Here you can find the most diverse variety of asteroids among all the world's temperate waters.

Like their spiny-skinned cousins in the other classes of Echinodermata, sea stars demonstrate some fascinating traits and abilities. Asteroid arms feature light-sensitive eyes; recent research indicates that in some species these eyes can distinguish shapes, and that asteroids will even position their arms to get a better "view" of what is around them. Like many echinoderms, sea stars rely more heavily on their sense of smell than sight, however—detecting predators, prey, and environments by sampling chemicals in the seawater that flows around and through them.

They have a simple nervous system but no brain; filtered seawater instead of blood; and a complex tissue-infused skeleton composed of individual components known as ossicles. These bony plates are connected with a handy and startling adaptation called catch connective tissue. When triggered by neural response, these tissues can stiffen or soften a sea star's external body wall, in some cases almost instantly. Using this nonmuscular method of becoming rigid and flexed is energy-efficient: intertidal asteroids like ochre stars can use it indefinitely to anchor themselves onto shoreline cliffs or boulders and resist the force of crashing waves. (Their tube feet also use a combination of suction and adhesive protein secretions to help them hold onto rocks and other surfaces.) When they want to change positions and move across rocks or the seafloor, they relax to become soft and flexible. Many species also

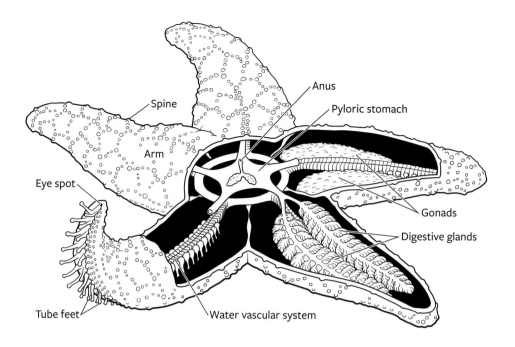

Spine

Anus

Pyloric stomach

Arm

Eye spot

Gonads

Digestive glands

Tube feet

Water vascular system

Anatomy of a sea star. Sea stars have a nervous system but no brain, filtered seawater instead of blood, and eye spots on their arms.

use their more rigid state to help them feed on prey, particularly difficult-to-open clams.

Sea stars themselves have few predators in the northeast Pacific. Sunflower stars and morning sun stars will attack other sea stars, and Alaska king crabs and sea otters have been known to feed on asteroids; when sea stars are attacked but not killed, however, they can demonstrate an amazing capacity for regeneration. An arm torn off or autotomously dropped in response to a predator can grow back on the main body—and even more impressive, the torn limb itself can grow into an entirely new sea star if it contains part of the star's central disk. The calcified bodies of ochre stars can be hard to break up, however, and to get a meal, hungry seagulls will sometimes choke them down whole.

Prior to the wasting disease epizootic, ochre stars—orange, purple, or somewhere in between—were a well-known fixture of the northeast Pacific coast. Easily spotted clinging underwater to pier pilings or fastened to shoreline boulders at low tide, they were familiar to just about anyone who explored these rocky intertidal zones. It may be from observing ochre stars that many of us get the impression that sea stars are fixed, sessile creatures, only slightly less stationary than barnacles. But asteroids display far more interesting behaviors than that. They are voracious predators, and many are quick on their tube feet.

The sunflower star, for example, is a roving lion of the benthic zone. The largest sea star in the world—it can have up to 26 arms and by some accounts measure a meter across—it glides across the seafloor hunting urchins, scallops, abalone, marine snails, sea cucumbers, and other prey at speeds of up to 10 centimeters a second. Sunflower stars are open-minded about their diet, which varies according to their specific location and includes sessile targets like barnacles and mussels. According to Drew Harvell, a professor of marine ecology at Cornell University, sunflower stars in captivity know when they are about to be fed and will begin climbing the sides of their tanks in preparation for a meal. She has seen them catch clams tossed into their aquariums "like a dog catches a ball."

Like most asteroids, sunflower stars have a feeding technique when eating bivalves and barnacles that is likely quite unpleasant for creatures on the receiving end: they can evert their stomach outside their body and—while pulling their prey's shell apart with their tube feet—insert it inside. Then the stomach itself surrounds the organism and digests its tissues in place. Their reputation as predators often precedes them in their native habitat. When the northern abalone (*Haliotis kamtschatkana*) senses the presence of a sunflower star, it speeds up—abalone are sea snails that use their large basal feet to move—and then, to break the grip of the sea star's tube feet, it will quickly whip its shell back and forth.

Hakai diver Kyle Hall checks out a sea star, showing its tube feet.

The northeast Pacific Ocean is home to half of the world's known sea star families. Many northeast Pacific species are found nowhere else on the planet. Top row, left to right: rose star (*Crossaster papposus*); juvenile sunflower star (*Pycnopodia helianthoides*); gunpowder star (*Gephyreaster swifti*); northern sun star (*Solaster endeca*). Center row, left to right: blood star (*Henricia* sp.); spiny red star (*Hippasteria phrygiana*); striped sun star (*Solaster stimpsoni*); cookie star (*Ceramaster* sp.). Bottom row, left to right: basket star (*Gorgonocephalus eucnemis*); slime star (*Pteraster tesselatus*); wrinkled star (*Pteraster militaris*).

Drew Harvell with a sunflower star at Friday Harbor Laboratories in Washington State.

Changing the seascape

Their skills as a top benthic predator, however, didn't help sunflower stars resist the wasting disease that nearly extinguished them. Sunflower stars act as keystone predators in some areas, and they play a major role in keeping down populations of sea urchins. Urchins feed on kelp, and their numbers have increased dramatically since the outbreak, with significant impacts on kelp forests—an important habitat in the northeast Pacific explored in the next chapter.

"As the largest epizootic of marine wildlife that's ever been documented, this was a landmark event," says Harvell. She notes that there have been larger, global-scale marine epizootics among farmed shrimp, but not among wild species, and never with such large ecological consequences. "What we've seen with sswd stands with other landmark disease outbreaks like white-nose bat syndrome or the chytrid fungus disease outbreak in frogs. It's been extremely destructive to biodiversity."

Harvell leads a National Science Foundation research network that created a brain trust of disease ecologists, pathologists, immunologists, and climate experts to investigate links between outbreaks in the ocean and climate change. Such events happen more frequently underwater than most of us might think—and investigations of disease in multiple marine species, and the ecological disruptions they cause, are laid out in Harvell's book *Ocean Outbreak: Confronting the Rising Tide of Marine Disease*. For example, California used to have the richest biodiversity of abalone species in the world, Harvell points out, but due in part to diseases, two of them are now on the endangered species list, and another three are listed as species of concern.

In the 1980s, the Caribbean black sea urchin (*Diadema antillarum*) was extirpated by a still-unknown pathogen in an event well known to marine disease ecologists. The epizootic affected urchins across approximately 3.5 million square kilometers of ocean, in some areas killing off 99 percent of them. Troubling outbreaks continue in the Caribbean, particularly among corals. The enduring presence of white band disease, black band disease, and stony coral tissue loss disease is having significant impacts on Caribbean species of reef-building corals. The Caribbean has only

8 percent of the world's coral reefs, but they are the site of 70 percent of reported disease outbreaks. This has added to the human-caused destruction in the region in recent decades.

"Disease can change our seascape," says Harvell. "Our reefs are melting away, not just from bleaching but also from infectious diseases. What we can say with a lot of confidence is that in some groups like corals, definitely the frequency of disease outbreaks has increased. We know the driver: it's ocean temperature."

Harvell was one of the first to begin exploring the causes of sswd. Previous outbreaks of the disease in the northeast Pacific—documented in California in the 1970s, through to an outbreak among ochre stars off Vancouver Island in British Columbia in 2008—were localized and never involved more than a single species. In some instances, these earlier outbreaks were found to correlate with warmer water temperatures, but the causal agent was not identified. Early in the epizootic that began in 2013, researchers speculated that starvation, oxygen depletion, or ocean acidification might be to blame.

Harvell and others in her network immediately began trying to identify the presence of an infectious pathogen, such as a bacterium, amoeba, or virus. Identifying such organisms in marine species, however, is notoriously difficult. "For disease pathologists who work on this kind of thing all the time, they know how hard it is to actually pin down an infectious agent," says Harvell. "It took over fifteen years to figure out the cause of the withering syndrome that was killing California abalone, and that was easier because it was actually a relatively large bacterium you can see under a microscope. Viruses are vastly harder and they do really weird stuff. They can get in, infect, and then essentially disappear."

As Harvell describes in her book, dissecting sea stars to search for potential sswd pathogens was a precise and delicate business. It meant dipping instruments in alcohol and then flame-sterilizing them to prevent cross-infection of the sample portions of sea star tissue, which were then preserved in formalin or flash-frozen at –80°c. Leading marine disease pathologists joined in—including microbiologist Ian Hewson and Bronx Zoo pathologist Harley Newton—to test the samples for viruses, bacteria, and other potential disease agents.

While scientists did find some evidence to indicate that the pathogen in question was a virus, or at least virus-size, as of 2019 the exact causal agent remains controversial. Research indicates that sswd affects the inner microbial communities in sea stars, and some researchers

speculate that the disease may be polymicrobial, caused by a suite of deadly microbes—as is the case with black band disease, one of the diseases afflicting Caribbean corals.

What is not in doubt, however, is that warming ocean temperatures amplified the effects of sswD. Multiple experiments confirmed that warmer waters made more sea stars get sick and caused them to die faster. As is discussed in more detail in Chapter 7, the onset of sswD coincided with the marine heat wave known as the Blob. A mass of warm water larger than Brazil, the Blob lingered in the northeast Pacific between 2013 and 2016, with suspected deadly impacts on salmon, sea otters, and fin whales, among many other species.

Using deep offshore trawl data and nearshore dive surveys from California to British Columbia, Harvell, Jenn Burt, Anne Salomon, Ondine Pontier, and other scientists found that the Blob impacted sea stars as well: sswD mortality in sunflower stars and ochre stars were correlated with warmer water temperatures. In California, where the waters are warmer, declines of the latter approached 99.9 percent in the intertidal zone; farther north in areas of the Central Coast around Calvert Island, the declines were in the range of 25 to 60 percent.

"The evidence is quite good now that temperature played an important role," says Harvell. "I think of it as temperature fueling an existing

outbreak and increasing the impact of it, rather than necessarily being causal."

Researchers are still working to understand the actual mechanism linking temperature and the disease—whether the sswd pathogen is more abundant in higher temperatures, or whether warm water simply made sea stars more stressed and less able to fight infection. In either case, warmer temperatures in the Pacific Ocean appear to be bad news for the many sea stars vulnerable to wasting disease.

Wasting disease continues to infect sea stars in areas throughout the northeast Pacific. Along with its causes, scientists are currently studying the effects that the massive epizootic has had, and continues to have, on coastal ecosystems. For community ecologists who study the relation-ships between organisms, the epizootic has created a rare chance to study large-scale shifts in species dominance. Diseases that remove or reduce the presence of certain species often mean that other animals fill in the gaps, using the habitat once dominated by the now missing organism and eating its former prey.

Christopher Harley returned to his Vancouver-area intertidal sites with researchers Alyssa-Lois Gehman and Sharon Kay to see what was occurring with asteroid species in the wake of the major sswd outbreak. Together they found that in some areas where populations of sunflower stars and ochre stars had plummeted, there was a corresponding boost in the numbers of less abundant competitor species like the leather star (*Dermasterias imbricata*) and the mottled star (*Evasterias troschelii*).

"The mottled star, for example, does suffer from this disease, but hap-pens to be a little bit less susceptible than the ochre star," says Harley. "So when ochre stars went down, the mottled star was also getting sick. But suddenly there was this ecological opportunity and we started seeing five to ten times more mottled stars at our field sites than we ever had before, because of this ecological niche that's opened up for them, with more available food and habitat."

Phenomena like these are part of observing potential post-sswd "regime shifts"—abrupt and possibly long-lasting changes in ecosystems—on the northeast Pacific coast. One of the most significant impacts of the sswd epizootic has been on kelp forests (see page 95). Healthy kelp beds create undersea habitats that are crucial to marine biodiversity, and, as researchers since Bob Paine have discovered, their existence in turn depends on keystone species.

Sea stars dying of wasting disease. Top, leather star (*Dermasterias imbricata*). Center, vermilion star (*Mediaster aequalis*). Bottom, striped sun star (*Solaster stimpsoni*).

5 FORESTS IN THE TIDE
SEA OTTERS, URCHINS, AND KELP

IT'S IRONIC THAT the undersea world is so alien to most of us, since it makes up over 70 percent of the globe. "Planet Earth," says marine ecologist Jenn Burt, "is actually Planet Ocean." Burt has years of experience exploring the ocean floor off British Columbia's Central Coast, but scuba diving still fills her with wonder.

"Most of us only ever see the surface of the ocean," she says. "When you descend below the surface, it's like going into space. You feel like you're on another planet, in a zero-gravity environment with all this life that you don't normally get to see."

Burt's graduate studies undertaken with the Hakai Institute were focused on the ecological and social impacts of sea otter recovery—including the effects that growing sea otter populations have on coastal Indigenous communities. Her mentors and colleagues include renowned coastal ecology experts Jane Watson and Anne Salomon.

Beginning in 2013, Burt and her colleagues studied rocky reef areas around Calvert Island, initially trying to better understand the effect that sea otters (*Enhydra lutris*) were having on undersea ecological communities. Part of a recovering North Pacific population that has expanded across the Central Coast over the past three decades, sea otters are a keystone species that indirectly influence the abundance of kelp forest habitat. Kelp forests in turn provide critical habitat and nutrients to a

wide variety of creatures, from the very small to the very large. Juvenile salmon, herring, and rockfish all use kelp beds for habitat, as do crabs, sea stars, abalone, and other sea snails. Seals are frequent visitors to kelp forests, and gray whales will feed on the crab larvae they find there—as well as use the kelp to hide their calves from killer whales.

After four years of research, Burt and her colleagues came to a groundbreaking new understanding of the relationships between kelp, urchins, and sea otters—thanks to unexpected discoveries about a previously overlooked bottom-dwelling predator.

What is a kelp forest?

Kelp is a type of marine algae. Also known as seaweed, there are red, green, and brown varieties of marine algae, among others. They can be as small as a human blood cell or, in the case of brown macroalgae called giant kelp (*Macrocystis pyrifera*), grow up to half a meter a day to reach lengths of over 45 meters.

Kelp forests grow along the rocky coasts of the northeast Pacific, from shallow intertidal areas to towering canopies in waters over 27 meters deep. In British Columbia and much of the northeast Pacific, they are generally composed of clusters of giant kelp and the whip-like bull kelp (*Nereocystis luetkeana*). In Alaska they can also be formed by dragon kelp (*Eualaria fistulosa*). Giant kelp is like a perennial plant, with a life cycle that can last up to seven years. Bull kelp grows and dies off within a single year, an impressive feat given that it can reach lengths of over 15 meters. Rather than using roots—as seagrasses do, which unlike algae are true plants—these graceful organisms cling to the rocky seafloor using growths called holdfasts.

Kelp are primary producers, which means they don't feed on other organisms, but produce their own food through photosynthesis—absorbing sunlight and organic carbon from the environment. Like land-based primary producers such as plants and trees, kelp and other algae form the base of the marine food web. They provide nourishment to grazers, the herbivores that feed on them. There are many grazers that feed on kelp; two examples are sea urchins and abalone. In turn, these grazers provide

PREVIOUS SPREAD: Baby rockfish in a kelp forest.

ABOVE: Nearshore researchers measuring bull kelp on Calvert Island.

food for predators, such as sea otters and certain species of sea stars like the sunflower star (*Pycnopodia helianthoides*).

A northern abalone making its way along the rocky seafloor. Like sea urchins, abalone—a type of sea snail—are voracious feeders on kelp.

Jane Watson is an authority on kelp ecosystems in British Columbia. "Perhaps the most important thing kelp does," Watson told a Hakai interviewer in 2017, "is increase the productivity of nearshore ecosystems. Kelp is essentially dumping carbon into that system, and it's doing that by leaching into the water and by breaking up into tiny little fragments which feed little animals we call detritivores. Those little animals go on to feed bigger animals, and in turn feed entire food webs."

Kelp are not just primary producers; kelp forests also provide a foundational habitat that is critical to many species. In a terrestrial forest, trees provide habitat for a vast range of life, from microorganisms and insects to large mammals. Similarly, the stipes and fronds of a kelp forest create a three-dimensional lattice that benefits everything from invertebrates like snails and anemones to marine mammals like seals and whales—providing food, hiding places, and shelter.

The biomass of kelp forests is enough to slow the movement of water, which offers sea life a buffer from surging storm waves. As a result, kelp provide a nursery function: as organisms like sea stars and crabs release

larvae, the slower water in kelp forests allows them to settle and develop rather than be swept away to a less hospitable location.

Like some seagrasses, kelp fronds provide direct surfaces on which herring can spawn. (Herring spawn can be so prolific that its weight can temporarily sink kelp forests.) By creating nearshore habitats rich in fish and other sea life—and by slowing down wave action and therefore erosion—kelp beds have been significant assets to human coastal communities for thousands of years. The abundance of food now known to be generated by kelp forests has influenced the rise of the kelp highway hypothesis. This idea holds that the first ice-age human populations to expand southward from Beringia (see Chapter 1) did so by exploiting the rich marine resources of ice-free coastal regions, rather than migrating via an inland corridor between ice sheets.

Kelp forests are influenced by a huge range of physical and biological factors. They are sensitive to temperature, ripped up by storms, and affected by currents and waves. Researchers have also discovered that kelp forests in the northeast Pacific can be dramatically impacted by the presence of two predators. One of them is cute and fuzzy—the other one, less so.

FACING: Top, bull kelp, like those shown here at the end of their life cycle, use gas-filled bulbs to float. Bottom, northern kelp crabs blend well into their environment, almost disappearing among the kelp stipes.

BELOW: Left, these graceful organisms are not jellyfish, but a type of sea slug called a hooded nudibranch. When out of the water, they give off a fragrance often compared to watermelon or grapefruit. Right, schooling black rockfish swim against the current in a kelp forest.

James Estes off Bering Island, Russia.

The sea otter connection

In 1970, James A. Estes traveled to Amchitka Island in Alaska's Aleutian archipelago to study sea otters. The Atomic Energy Commission sponsored his research there ahead of a contentious nuclear test on the island. Despite the Cold War considerations that placed Estes in the Aleutians, his research there ended up transforming our understanding of coastal ecology. He would return there to research sea otters and kelp forests for decades afterward, as he describes in his 2016 memoir, *Serendipity*.

Observing sea otters and kelp beds on Amchitka—both onshore and during scuba dives—led Estes to question the links between them. The crucial connection, he discovered, was sea urchins. Sea urchins are voracious grazers of kelp, and sea otters, in turn, are equally voracious predators of sea urchins. But Estes didn't understand the real meaning of what he'd seen until a meeting with ecologist Bob Paine (see Chapter 4) on Amchitka in 1971. Paine's pioneering research with sea stars had defined the role of keystone species and the phenomenon he called a trophic cascade—the knock-on impacts on a food web when a top predator is removed.

As Estes tells it in his autobiography, he was "intimidated" at the thought of meeting Paine, but Paine was undeterred. He sought Estes out for a conversation one evening, and the event was a turning point for Estes. "Bob listened to my account of what I had seen while diving and what I thought it might mean and then abruptly suggested a simple but radically different change in perspective. Rather than wondering how the kelp forests affected otters... why not explore how the otters affected the kelp forests?"

On Amchitka, Estes's observations revealed an "extraordinarily high" sea otter population and an undersea world of lush kelp forests. To understand the relationship between the two, Estes realized he would have to emulate the approach Paine had taken when exploring the effect of ochre stars (*Pisaster ochraceus*) on intertidal ecology. As described in the previous chapter, Paine had simply removed ochre stars from the rocks and tossed them into the sea. Removing sea otters from the waters around Amchitka was out of the question, so Estes had to try a different strategy: he needed to find an area that was devoid of otters and see what the kelp forests looked like.

Estes was able to find just that about 350 kilometers to the west of Amchitka, off a remote island called Shemya. As he tells in *Serendipity*, what he saw when he put on his scuba gear and entered the waters off Shemya shocked him.

> When I looked down at the seafloor, I was stunned by the vast numbers of urchins and the absence of kelp... Every place I looked was the same—large and abundant sea urchins over a seafloor of crustose coralline algae with little or no kelp... In the absence of sea otter predation, sea urchins had increased in size and number, and the larger and more abundant urchins had eaten the kelp. This was my "aha moment," a profound realization that would set a path for the remainder of my life.

Estes had seen what is now known as an urchin barren, the result of a trophic cascade. He would go on to realize that—like Paine's ochre stars on intertidal rocks—sea otters are a keystone predator that increases the abundance of a diverse array of sea life. In coastal areas where sea otters regularly consume sea urchins, kelp forests have a greater chance to take hold and endure, and act as an undersea forest habitat for other

A carpet of green, purple, and red sea urchins turning a kelp-bed ecosystem into an urchin barren.

marine animals. But when there are not enough predators like sea otters to feed on sea urchins, the urchins graze over and effectively wipe out kelp forests.

Thanks to Estes and the work of subsequent researchers like David Duggins and Jane Watson, the beneficial effects of sea otters in propagating kelp habitats have been well established among ecologists, and increasingly well known to the general public. However, a new twist on this relationship—usually understood as a three-way interaction between urchins, kelp, and sea otters—has recently come out of the work done on the Central Coast by Jenn Burt and her Hakai Institute colleagues.

Unnatural experiment

In the waters around Calvert Island, where surface temperatures dip as low as 8°C, subaquatic researchers need an airtight dry suit along with their usual scuba gear. Burt and her colleagues also carry a panoply of marine science tools: a foldable, two-dimensional box for sampling called a quadrat; sealable bags for collecting everything from kelp to invertebrates; a writing slate with data sheets, pencil, and built-in ruler; and meter tapes for measuring survey areas, known as transects.

"By the time you clip all the gear onto yourself so that you have everything and it doesn't float away," Burt says with a laugh, "it can look like a real gong show."

For Hakai Institute scientists, undersea surveys bring other challenges. Rough weather on the outer coast can keep research vessels tied up at the dock for days. When boats are able to go out to gather data—in Burt's case, counting sunflower stars, or performing an undersea experiment to track urchin feeding rates—divers have only 45 to 60 minutes to complete their task before their oxygen runs out. The need to stick with your "dive buddy" and communicate with hand signals or by scrawling on a slate can slow things down, as can murky water. Perhaps the biggest challenge is that the waters of the Central Coast are rarely still. Unlike the woods on land, a kelp forest moves with the push and pull of the ocean—and so do the researchers. While the swaying kelp stipes of what Burt calls "liquid forests" can be serene and graceful, at other times wave and tidal currents can be more dramatic. Burt recalls counting urchins and other

FACING: Top, a raft of otters on the Central Coast. Bottom, a sea otter in a kelp bed.

BELOW: Jenn Burt geared up for a research dive.

IS IT A SEA OTTER?

For most of us—at least those who don't live in Monterey, California, which has a stable and fairly visible population of about 2,800 sea otters in the outlying bay—seeing a sea otter (*Enhydra lutris*) is a rarity outside of aquariums. River otters (*Lontra canadensis*) are often mistaken for sea otters because they are just as happy to forage in the sea as they are in freshwater lakes and rivers. Both belong to the weasel family, Mustelidae, a group of carnivores that also includes badgers, ferrets, and wolverines. Among its fellow mustelids, the sea otter is the only purely ocean-dwelling species. One of the best ways to tell river otters from sea otters is their tails. A sea otter's tail is about one-third the length of its body; a river otter's tail is much longer, about two-thirds its body length. River otters scamper around quite effectively on land, feeding and making their dens there, but sea otters rarely spend time ashore; they've perfected the art of floating on their backs in the sea, eating and even sleeping in that position. (River otters, by contrast, are not known to swim on their backs.) Mature sea otters are also larger: river otters top out at about 14 kilograms, but sea otters can weigh up to 45 kilograms. ■

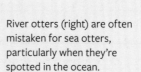

River otters (right) are often mistaken for sea otters, particularly when they're spotted in the ocean.

invertebrates with Ondine Pontier, now lead kelp researcher at the Hakai Institute, in a "sloshy zone of surge" near Calvert Island.

"You'd swim up to your quadrat on the seabed, count a few things, and then the surge would suck you back six feet. Then you'd swim back, count a few more things, and then you'd be pulled back again, over and over," she says. "It would have looked pretty comical if you saw it in a video."

Enduring the challenges of underwater experimentation and monitoring, however, can lead to groundbreaking insights. Burt and her fellow researchers began collecting data in 2013, not knowing that two years later, the epizootic of sea star wasting disease (SSWD) would strike the waters around Calvert Island. (See Chapter 4 for more on the disease outbreak.) The timing turned out to be serendipitous.

Burt's original focus was to track ecological communities before, during, and after the arrival of sea otters, which—because of their post–fur trade population recovery (see page 92)—are still recolonizing reef habitats across the North Pacific Rim. The waters around the Hakai Institute are at the leading edge of this expansion, and like the Aleutian Islands, are well suited to this sort of study. There are areas with and without sea otters where researchers can compare the differences in kelp abundance, urchin populations, and other aspects of undersea habitat. When SSWD struck the region during the winter of 2015, it suddenly brought a new dimension to the research: a rare opportunity to also look at how predatory sea stars influence coastal rocky reef ecosystems.

"This is the important element of surprise in science," says Burt. "We just don't always know what we're going to find."

The disease outbreak created what is known to researchers as a natural experiment—a unique condition that appears without artificial or planned intervention. By wiping out huge numbers of sea stars, including those in nearshore kelp-producing zones, SSWD had resulted in a vast, multispecies version of Paine's ochre star elimination experiments.

"There are so many interactions in the ocean that we don't totally understand and probably never will, because we can't systematically remove and then replace all the animals in the ocean," says Burt. "But when we have that opportunity to observe a sudden absence of a particular organism, and see what the impacts of that are, we have a lot more understanding. That's the bigger narrative around this research that I think is really neat."

Using a dataset collected over four years, Burt and her colleagues learned that sea otters are not the only predators that have a significant effect on urchins, and thus not the only creature that can help ensure the health and abundance of kelp forests: the less cute but nonetheless impressive sunflower star also plays a major role. On the Central Coast reefs she studied, Burt links the loss of sunflower stars to a startling 300 percent spike in the population of small and medium-size urchins, and to a resulting 30 percent decline in kelp density.

What Burt noticed on underwater surveys after the disease event was that there were more small and medium-size urchins everywhere Burt and her colleagues looked. That gave the team some hunches about what they wanted to look at more closely in the data. When they did, says Burt, the insights came not only from her, but from "looking at the system with some clever modelers, from having coffees with Jim Estes, and from talking with people who had seen similar things in other areas with green urchins."

Sea otters do an excellent job of eating large sea urchins over eight centimeters in diameter, but they tend to leave behind smaller urchins. These small urchins also readily devour kelp, and if their numbers swell, they begin to thin out kelp forests. However, these smaller urchins are a favorite food of sunflower stars. "Sunflower stars mop up the little sea urchins," says Anne Salomon. "It was very interesting to learn how they give a helping hand to the apex predator in these areas, sea otters."

Salomon notes that although many ecologists had suspected that sunflower stars played a role in the resilience of kelp forests, that role had not yet been measured or acknowledged. "Sea star wasting disease came and did the experiment for us, and research on the Central Coast with Jenn Burt provided evidence of the important role that sunflower stars have."

"It really is like that Joni Mitchell line, 'you don't know what you've got till it's gone,'" says Burt. "When you remove an organism from the system, it's then you realize that its impact was actually quite substantial."

Researchers from British Columbia to California have seen large swaths of kelp forest disappear off the coastline in recent years, a troubling development that appears connected both to rising ocean temperatures and sswd. Confirming the previously uncertain role of sunflower stars in kelp-bed ecology will help researchers as they monitor the resilience of kelp beds across the North Pacific Rim and manage their recovery. The local presence or absence of sunflower stars may also help explain why kelp beds have recovered at different rates in areas newly colonized by sea otters: kelp beds with low populations of sunflower stars may propagate more slowly.

Scientific insights are built on collaboration. Surveys by Watson, Erin Foster, and sea otter researcher Linda Nichol provided a key piece of the puzzle, documenting the otter population and its seasonal use of local areas. Burt's research also received a big boost from what Watson calls an "epic" effort to generate maps of kelp beds in British Columbia using satellite imagery and drones. This sort of mapping is more difficult in British Columbia than it is in California, due to a variety of factors—including cloud cover and seasonal storms, a more complex coastline of islands and fjords, and, recently, the obscuring smoke from forest fires. Spearheaded by Hakai researchers Keith Holmes and Luba Reshitnyk, this initiative to piece together a bird's-eye view of kelp beds was paired with ongoing "fish-eye view" scuba surveys by Pontier and other scientists at the institute.

"It's that emerging narrative that comes out of collaboration and discovery, and being in the right place at the right time," says Burt, "all of it coming together to illuminate new patterns, interactions and understandings."

LEFT: Anne Salomon, left, and Brooke Davis sampling intertidal kelp on Calvert Island.

RIGHT: Jenn Burt surveying urchins.

ABOVE: Winter otter surveys off Calvert Island.

RIGHT: Sea otter dining on a sea cucumber.

FACING: Top, sea otters, like this one dining on turban snails, frequently use stones to break open the shells of their prey. Bottom, a broken urchin shows the rich roe prized by otters and humans.

DISCOVERING DIET DIVERSITY

Sea otters can dive to depths of up to 300 feet to find food items such as urchins, abalone, clams, crabs, and other shellfish. Their habit of using rocks to break these morsels open—while floating on their backs, or with stationary "anvil" rocks—makes them the only marine mammal to use stone tools, and places them among a larger but still select group of animals who use tools for foraging and other tasks.

A five-year study led by Hakai researchers Erin Foster and Leah Honka—in collaboration with Linda Nichol and other sea otter researchers in British Columbia and beyond—recently discovered that the diets of sea otters in the coastal waters of British Columbia change over time. They found that when sea otters first arrived at an area near Calvert Island called Choked Passage, their diets consisted almost exclusively of large sea urchins. Urchin shells are called tests, and researchers found that sea otters preferred urchins with tests measuring 10 to 18 centimeters in diameter. (The largest of these urchins were up to 200 years old, alive since the mid-1800s when sea otters were extirpated from the Central Coast.)

However, within months, the sea otters at Choked Passage moved from foraging over rocky reefs to soft sediment areas of the seafloor, where they mostly ate clams. The switch to clams started once the otters had depleted the large urchins. Sea otters need lots of caloric energy in their diets, and they don't waste time hunting for large urchins once they become rare—and smaller urchins less than 10 centimeters in diameter rarely have enough edible mass to make them worth the trouble. In a nearby area that sea otters had occupied for two

years, their diet diversified to include kelp crabs, clams, abalone, and sea cucumbers—while in areas where sea otters had been foraging for between 3 and 30 years, their diet consisted mainly of small clams, kelp crabs, and mussels.

New insights about sea otter foraging habits will help inform the way human communities respond to their presence and support their resurgence. Foster and Honka's research team has also noted that the foraging habits of male sea otters in these areas differed from females. Male sea otters primarily sought food in deep channels and sheltered areas; females, who need to keep watch on pups, generally preferred shallower depths.

Observations like these provide vital context for ongoing sea otter research and management, says coastal ecologist Jane Watson. "For people involved in carnivore conservation, knowing that the needs of a recovering species may change with time, and differ between sexes, is crucial." ■

THE TOLL OF THE
MARITIME FUR TRADE

Sea otters (*Enhydra lutris*) were once an abundant species along the coastlines of the North Pacific Rim. Experts have estimated their historical population at about 300,000, occupying shoreline habitats in a broad arc stretching from Japan and Russia over to Alaska, and south to Baja California.

That changed following a 1741 expedition by Danish explorer Vitus Bering. The expedition's survey of southwest Alaska on behalf of the Russian Empire ended in a shipwreck and in the deaths of Bering and almost half his crew due to scurvy and other causes. The survivors returned with a cargo of 900 valuable sea otter pelts, however, and tales of the abundant populations of sea otters they had witnessed in the Aleutian Islands and beyond. This would ultimately kick off a "fur rush" for sea otter pelts, and a maritime fur trade that lasted for over 150 years.

Human interest in sea otter pelts was due to a unique evolutionary adaptation. Unlike seals and other marine mammals, sea otters do not have blubber to keep them warm in cold Pacific waters. Instead, sea otters have high metabolic rates (spending most of their waking hours feeding) and the densest fur of any mammal on Earth, with approximately 600,000 to 1 million hairs per square inch. In comparison, humans have less than 150,000 hairs on their entire head. Wealthy buyers in China, Russia, and Europe prized the plush fur, and in the 1840s a single otter fur could fetch as much as nine times the value of a beaver pelt.

Like primates, otters are social creatures. Mothers tend to stay with pups, whereas males will congregate in large groups called "rafts" as far as 150 kilometers away from females when not breeding. These sexually segregated rafts can number from three to a thousand individuals. Larger rafts would have made easy-to-locate targets for the many hunters that dramatically reduced sea otter populations during the maritime fur trade.

By 1911—when international agreements began to manage fur seal and sea otter populations, at least in some areas of the North Pacific Rim—sea otters had been brought close to extinction. Across their former range, only an estimated 2,000 individuals remained, less than 1 percent of their original population. They were clustered in close to a dozen remnant groups dotting mostly remote areas in Russia and Alaska. Sea otter populations have rebounded, but in many places they have not yet returned to historic levels.

Current estimates place the global population of sea otters at about 125,000, with about 89,000 of these in Alaska. The BC population is about 8,000—most of whom descend from a population of 89 otters that were relocated from Alaska to Checleset Bay on the west coast of Vancouver Island over three years beginning in 1969. Across most of their current range, the sea otter population continues to grow, but some areas have seen severe reductions. In Russia's Kuril Islands and Kamchatka Peninsula, the number of sea otters has declined up to 50 percent in some locations, possibly due in part to poaching for black-market pelt sales. In Alaska's Aleutian Islands, sea otters have been in decline since the 1980s, possibly linked to predation by killer whales or sharks. Citing losses in the past three decades

Sea otters in Choked
Passage near Calvert Island.

due to oil spills, predation, disease, and other factors, the International Union for Conservation of Nature has placed sea otters on its Red List as an endangered species.

Burgeoning local sea otter populations, however, have caused challenges in some coastal communities. Sea otters are voracious eaters, consuming up to a third of their body weight in a single day. This can put pressure on shellfish populations—urchins, abalone, clams, crabs, mussels—that humans also harvest. As the Nuu-chah-nulth Tribal Council's fisheries agency, Uu-a-thluk, wrote in 2011:

...now we find that sea otters are once again playing a large role in a shifting Nuu-chah-nulth society, as we see the impacts that their increased presence is having in the nearshore marine environment. Nuu-chah-nulth are challenged to ask themselves difficult questions about the economic, social, and spiritual impacts sea otters are having on their communities. ■

ABOVE: A rocky urchin barren on the Central Coast.

RIGHT: The underside view of an urchin, showing the dental structure known as Aristotle's lantern, which it uses to scrape algae and consume kelp.

FACING: Left to right, the red sea urchin, green sea urchin, and purple sea urchin are the most common urchin species in the waters of coastal British Columbia.

ZOMBIE URCHINS

In 2019, the Oregon Department of Fish and Wildlife conducted sea urchin surveys on a 40-square-kilometer reef on the state's southern coast. Their calculations estimated that the reef was overrun with 350 million purple sea urchins (*Strongylocentrotus purpuratus*)—a hundred-fold boom in the population in only five years.

Purple urchins are one of several species of urchins found off the Pacific coast of North America—including the commercially valuable red urchin, and green, white, and crowned urchins, among others. Scientists have linked the increasing abundance of urchins in California and southern Oregon with warmer waters from a marine heat wave (see page 120) and the near-total disappearance of one of their major predators, the sunflower star (*Pycnopodia helianthoides*), due to the outbreak of sea star wasting disease (see Chapter 4).

Urchins are voracious grazers of kelp, and in some areas the impacts of their population boom have been dramatic. Beginning in 2014, bull kelp coverage over a 350-kilometer stretch of coastline in Northern California was reduced by more than 90 percent in only one year. While smoking-gun causality is never easy to establish in the ocean or in biological science generally, the loss of kelp forests—a catastrophic loss to a wide variety of species that depend on them—appears directly connected to the spike in urchin numbers. The destruction in the region triggered the shutdown of a recreational fishery for red abalone (*Haliotis rufescens*) and a commercial fishery for red sea urchin (*Mesocentrotus franciscanus*), both valued in the millions of dollars.

Shellfish researchers with the Oregon Department of Fish and Wildlife noted that there wasn't enough food for all these new urchins on the reef, and that they were starving. A common-sense understanding of ecology might lead one to think this would restore ecological balance, as populations shrink to suit available resources—but in this case, that would be incorrect.

Enter: the zombie urchin.

Urchins have a remarkable ability to resist death by starvation. By using nutrients stored in their reproductive organs—which are a prized seafood in many Pacific cultures—and subsisting in a dormant state on whatever nutrients they can find, armies of these "zombie" urchins can persist for unknown lengths of time, hunkering down in what is known as an urchin barren.

"When animals are deprived of food, they stop any functions that are not absolutely necessary for survival," says marine ecologist Lynn Lee, who has studied sea urchins in British Columbia for over 25 years. "Gonad production is one of those

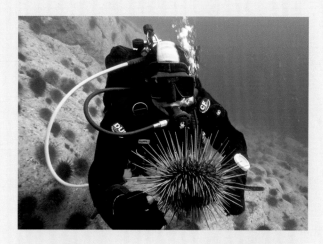

TOP: Daniel Okamoto with a red sea urchin.

BOTTOM: Lynn Lee conducting sea urchin experiments in Haida Gwaii.

things. If you don't have enough food, you stop reproducing."

Daniel Okamoto, a marine ecologist in the department of biological sciences at Florida State University, agrees. Okamoto has worked with Lee and with the Hakai Institute in studying sea urchins. "If you or I didn't eat for a week, we'd be in pretty rough shape," he says, "whereas urchins can do incredibly well."

Sea otters will forage from urchin barrens, but because most of the urchins in these denuded areas have very little soft tissue remaining inside them, they appear to be less appealing to other predators. This applies not only to humans that might harvest them for the rich roe known in Japan as uni; some experiments indicate that even spiny lobsters in California can detect which urchins are zombies, and that given a chance will choose urchins with larger reserves of roe. This has the potential of creating a feedback loop, where overabundant urchins are less likely to endure the predation that might reduce their numbers—and in some areas make urchin barrens a resilient, alternative state that resists the return to a kelp-bed ecosystem.

Lee now works as a researcher in the National Marine Conservation Area Reserve in Haida Gwaii off British Columbia's northwest coast. She and Okamoto have studied urchin growth rates in the waters around Haida Gwaii, tagging urchins and returning a year later to recover them and measure their size. They found that even in apparently foodless barrens, most urchins were growing: compared to urchins with access to kelp, large urchins in barrens grew more slowly, but small urchins grew just as fast.

Urchins in barrens can survive, says Okamoto, by "scratching away" a living eating whatever they can

Wiigaanad Sid Crosby
sampling urchin roe in
Haida Gwaii.

find—tube worms, coralline algae, drift kelp, even each other—and by altering their respiration. Using aquarium tanks with oxygen sensors at laboratories in Gwaii Haanas National Park Reserve and on Quadra Island, Okamoto and his PhD student Nathan Spindel found that urchins from barrens respire differently than urchins well fed on kelp.

"It's as if they've basically shut down. Their respiration rates are dramatically reduced," says Okamoto, noting that while urchins in barrens appear dormant, they have no problem waking up for a meal. "If their metabolic rate is way down, in theory they're going to have a much smaller demand for energy and are going to consume less. But what we're finding is that, no—actually, if you feed a starved zombie urchin, if you give it an opportunity to eat something, it's going to chow down."

How zombie urchins accomplish these metabolic feats and the length of time they can live in barrens are still mysterious, says Lee. "Urchin fishers and others have known about barrens for a long time. But the mechanisms behind this zombie quality, and how long an urchin can stay in this state with very little food and still survive, those are still very large unknowns."

Since the 1970s, marine ecologists have witnessed that when sea otters move into urchin barren habitats, they restore kelp forests (see page 83). But while the sea otter population is on the rise in British Columbia, in recent years it has declined or remained relatively low in other areas of North America's Pacific coast. Sea otters have yet to make a post–fur trade comeback in Haida Gwaii, so Lee is working on a cooperative project—involving the Haida Nation, Parks Canada, Fisheries and Oceans Canada, and the Pacific Urchin Harvesters Association—to attempt to restore kelp forests. Divers with the Haida Fisheries Program and commercial urchin fishers are removing urchins for food and culling zombie urchins in barrens over a three-kilometer stretch of coastline.

Lee has a key role to play in monitoring the project, but she is quick to remind people that urchins are not in themselves a wholly destructive force. "Urchins are not these nasty, terrible creatures all around," Lee says. "They're grazing and eating things and turning them into other materials that work into food webs. They're very important in the ecosystem. But when they're hyperabundant because their natural predators are missing, that's when there can be a problem." ∎

6 SEAGRASS
CRITICAL HABITAT

A Conversation with Emmett Duffy

A marine biologist who has worked in seagrass and coral reef ecosystems around the world, Emmett Duffy is director of the Smithsonian Institution's Marine Global Earth Observatory (MarineGEO) project. MarineGEO is a global partnership researching how coastal ecosystems work and how to keep them working. Author of the forthcoming book *Ocean Ecology: Marine Life in the Age of Humans*, Duffy is also founder of the Zostera Experimental Network, which studies eelgrass, among the world's most widespread marine plants.

Q: Why are seagrass habitats important?

A: All over the world, eelgrass and other seagrasses are foundation species that provide nursery and critical habitat for thousands of types of fish, crustaceans, and other marine life. Some tropical ocean grazers, such as manatees, dugongs, and green sea turtles, depend on seagrass directly as food. An adult dugong can eat up to a 10th of its body weight, about 40 kilograms, of seagrass in a single day.

Seagrasses are amazing plants. Eelgrass (*Zostera marina*) can live essentially forever as far as we can tell, by shooting out year after year from these rhizome mats, the roots that they have below the sediment.

PREVIOUS SPREAD:
Phyllospadix surfgrass, a
distinct species from eelgrass, on Calvert Island.

ABOVE: Left, Emmett Duffy
in the field in Panama. Right,
Duffy wading in the eelgrass
shallows at the Calvert Island
Ecological Observatory.

Twenty years ago, researchers working in the Baltic Sea genetically tested eelgrass and found that this one patch over 6,000 meters square was genetically identical—in other words, that this huge bed of seagrass was a single organism. They believed it was a thousand years old and the largest marine plant in the world ever measured.

You'll find eelgrass growing in the Pacific and the Atlantic, growing north to the Arctic from the waters of Korea, Mexico, North Carolina, and Portugal. One of its bizarre qualities is how different it can appear in different areas. In northern Japan, *Z. marina* eelgrass is up to three meters tall, with sparse but fairly big, thick shoots, almost like kelp. In Chesapeake Bay in Virginia, eelgrass gets tall in June, but through much of the year it's about a foot tall and incredibly dense, with hundreds of shoots per square meter, much more like a lawn.

Q: What are some of the issues facing seagrass habitats?

A: Seagrasses have been threatened and declining all over the world for a long time now. Some estimates suggest we've lost 35 percent of seagrass meadows globally since the early 1980s. This has happened largely under the radar, because unlike coral reefs, until recently not many people paid attention to seagrass habitats. In some areas, in fact, they have been regarded as a nuisance: people don't want eelgrass washing up on their beach, and so on.

The classical paradigm is that seagrass decline is caused by poor water quality from human-produced sewage, effluent, and agricultural runoff. There's no question that that's true. Sediments and increased nutrients create algae blooms; these can harm eelgrass, for instance, by shading it from the sunlight it needs to grow, or by growing directly on it. However, more and more evidence has been mounting in the last decade that food-web disruptions can also cause significant negative effects. In the Baltic Sea, over many years some smart ecologists have pieced together that overfishing of cod in the open Baltic has cascaded down the food chain to cause blooms of nuisance algae—with resulting declines of seagrass in the coastal areas.

Superficially this looks like nutrient pollution because you have a lot of algae growing on the seagrass, but what's really happening is that reducing the cod population allows the small fish that they prey on to increase. Those small fish in turn eat the mesograzers, these little herbivorous crustaceans and snails that live on the grass and keep it clean of algae that would otherwise overgrow it. And so you have this knock-on effect all the way down that allows explosions of this nuisance algae that knock out the grass. It's similar to the sea otter story: you've got a top

Female kelp greenling in eelgrass.

Striped seaperch in Choked Passage off Calvert Island. Seaperch are one of the few fish that are viviparous, giving birth to live young.

predator, and when you take that away, the population of its prey, which is the next trophic level down, explodes. That animal then puts tremendous feeding pressure on the organisms at the next trophic level.

Q: Is climate change affecting seagrasses?

A: Marine heat waves are definitely part of the picture, and these are linked to global warming. Shark Bay in Australia is a World Heritage site that has 12 types of seagrass out of the more than 70 species in the world. Seagrass covers about 4,000 square kilometers of the seafloor there, and it's a tremendous ecosystem for tiger sharks and sea turtles and all kinds of creatures. But in 2011 they had a big die-off of seagrass, with losses up to 90 percent at some sites. That was linked to flash floods that delivered a lot of sediment into the area—which were themselves a product of climate change—and to a warming event in the ocean.

Q: A main focus of current eelgrass work, which grew out of the Zostera Experimental Network, is eelgrass wasting disease. What is it?

A: We often think of keystone species as being animals like sea otters and sea stars. The definition of a keystone is a species that has an impact

SEAWEED OR SEAGRASS?

Is eelgrass a kind of seaweed? Is kelp a kind of seagrass? The answer to both of these questions is... no. Kelp is a type of seaweed, and eelgrass is a type of seagrass. Both can create nutrients from sunlight using photosynthesis, but they are not the same kind of organism. So what sets them apart?

Seaweed is another name for marine macroalgae. They don't have flowers or seeds, and they absorb their nutrients across the surface of the plant rather than only through true roots. Many macroalgae have a root-like mass called a holdfast that anchors them to the seafloor (see below, right).

Seagrasses, by contrast, are true plants whose closest ancestors are terrestrial grasses. Like them, seagrasses have flowers, seeds, and vascular tissues called xylem and phloem that carry nutrients from their roots throughout the plant. They even produce pollen—the longest pollen grains in the world, up to 50 times longer than typical terrestrial varieties.

Examples of seagrass are surfgrass and eelgrass, with eelgrass being the most common in the northeast Pacific. ■

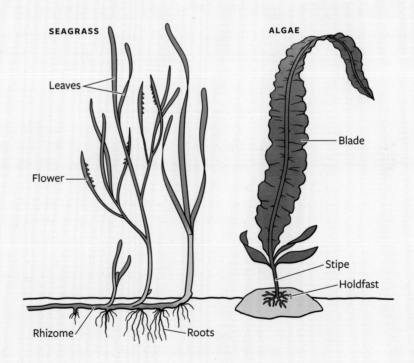

SEAGRASS

Leaves

Flower

Rhizome

Roots

ALGAE

Blade

Stipe

Holdfast

LEFT: Gillian Sadlier-Brown on a seagrass survey.

RIGHT: A diver examining herring spawn on eelgrass.

disproportionate to its abundance, or biomass: organisms that have a big impact even though there aren't very many of them around. By that definition, a pathogen that causes disease is the ultimate keystone, because it has virtually no biomass at all and can cause catastrophic damage that changes entire regional ecosystems.

That was seen, or at least strongly inferred, in the Atlantic back in the 1930s. Historically, eelgrass was found all along the temperate coasts of the entire Northern Hemisphere—pretty much everywhere that there was suitable habitat for it, which is sandy or muddy sedimentary shorelines where the waters aren't too rough. Then in the 1930s, there was a massive eelgrass die-off on both sides of the North Atlantic. It hit European waters hard, and killed almost 90 percent of the eelgrass on the east coast of North America.

At the time, people had no idea what was going on. Eelgrass is an incredible organism that was dominant throughout the coastal regions of the North Atlantic, and the die-off had big impacts because there were, and are, a lot of fishery species that live in eelgrass habitats. In those days people even used dried eelgrass directly for all kinds of things, including stuffing mattresses and upholstered furniture.

TOP: Anchovies schooling in an eelgrass bed off Vancouver Island.

BOTTOM: Harbor seals in a nearshore surfgrass and kelp habitat.

The main culprit in wiping out eelgrass in the 1930s wasn't discovered until much later—until about 20 years ago in fact, which at my age now seems like not all that long ago. While there may have been other contributing factors, the primary cause turned out to have been a wasting disease caused by a protist, *Labyrinthula zosterae*. It's a kind of slime mold that gets inside the tissues of the grass and slowly wastes it away. Its biology is not well understood, which was a major motivation for our current eelgrass disease project. We wanted to organize researchers to look at a variety of ecological questions relating to eelgrass, including this disease, which we just call Labby.

Q: Is Labby a big threat to eelgrass now?

A: There appear to be lots of places around the world where Labby is living on eelgrass and the eelgrass beds are doing just fine. So the situation right now is not epidemic-level disease. But every once in a while you get a situation where the disease takes off and meadows of seagrass start dying. It's unclear what the trigger for that is.

I'm not an expert in marine disease like Drew Harvell or many others, but from my admittedly limited knowledge, what we've seen with Labby seems typical of marine diseases: they have complex causality. Oftentimes we really don't know what causes them to take off. So with some partners we are exploring eelgrass habitats from San Diego to Alaska in an attempt to understand the holistic ecology of disease.

It's a tremendous group, including researchers from the Hakai Institute and universities in Alaska, Washington, Oregon, and California, with coordination by the Smithsonian Institution. Cornell, UC Davis, and the University of Central Florida are also key partners, leading work from drone technology to microbiome research to computational biology.

We're looking at a variety of factors. There is some evidence that warmer waters can kick off an outbreak, so there are lots of implications for climate change. Seagrass that has algae growing on it can be stressed, and that also might make them vulnerable to disease. We're tracking temperature, the algae that are growing on the grass, the fish and other animals that are wandering around in the grass, and the mesograzers that feed on algae. We're also mapping seagrass beds with aerial drones in the hope that if we do see the disease at ground level, we can follow that over time and see if it actually has an impact at the landscape level.

FACING: Top, hooded nudibranchs and juvenile rockfish in eelgrass habitat. Bottom, eelgrass researchers in Pruth Bay on Calvert Island.

OTTERS AND SEAGRASS: BETTER TOGETHER?

Elkhorn Slough is a salt marsh in Monterey Bay, California, flanked by thousands of acres of agricultural fields. Marine biologist and Sonoma State University assistant professor Brent Hughes has called it "one of the most nutrient-polluted estuaries on the planet."

Excesses of nitrogen or phosphorus from agricultural fertilizers can cause algae to become overabundant in streams, lakes, and estuaries, reducing oxygen and sunlight in the water—a process called eutrophication. Thick algal mats can even form on the surface, creating a suffocating green carpet. In waters like these, algae will grow on the blades of the seagrass and block the marine plants

from receiving sunlight, hampering growth or even killing them.

This means that in Elkhorn Slough, seagrass shouldn't survive—but it does. The reason, Hughes and his partners have discovered, is a three-way relationship between sea otters, crabs, and types of small grazers that clean algal growth from eelgrass. The trophic cascade (see page 60) works like this: along with small isopod crustaceans (*Idotea resecata*), a type of sea slug called an eelgrass sea hare (*Phyllaplysia taylori*) eats the algae that grow on eelgrass. Rock crabs (*Romaleon antennarium* and *Cancer productus*) in the estuary eat these little crustaceans and sea hares. If sea otters weren't

FACING: Sea otters in Elkhorn Slough in California.

LEFT: Eelgrass sea hare feeding on epiphytes.

RIGHT: Sea otter feasting on a clam.

around, crabs would eat the little grazers and algae would overgrow and kill the eelgrass. But the presence of over 120 hungry sea otters in the slough keeps the rock crab population down and allows an abundant population of grazers to clean the eelgrass and keep it healthy.

Sea otters colonized Elkhorn Slough in 1984. Impressed by the role of otters in this estuary habitat and their continued success in inhabiting it—a phenomenon that has challenged traditional ideas of sea otters as thriving only in kelp forest ecosystems—Hughes and other researchers have suggested that California could potentially triple its population of approximately 3,000 sea otters by introducing them directly into San Francisco Bay. Historically abundant in San Francisco Bay prior to the maritime fur trade (see page 92), sea otters on their own have not yet been able to recolonize the bay, the largest estuary on the North American west coast, because of the population of great white sharks that congregate near its narrow mouth at the Golden Gate Bridge.

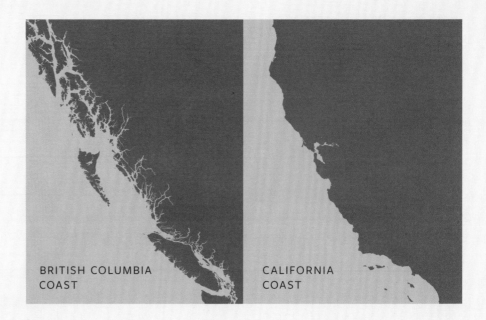

The complex glaciated BC coastline offers an abundant and varied habitat of fjords and islands. California, by comparison, has much less coastal habitat for the same range of latitude.

BRITISH COLUMBIA
COAST

CALIFORNIA
COAST

Digging into genetic biodiversity

Farther north in British Columbia, sea otters have more options when choosing places to forage (and, in most locations, less to fear from sharks). Part of this is due to geography. The stretch of British Columbia's coast that currently hosts sea otters is similar to California in terms of latitude range; according to analyses by marine ecologist Margot Hessing-Lewis and others, however, within that zone—because of its complex coastal geography of islands, bays, and fjords—British Columbia has 10 times as much shoreline. The coastal habitat differs in other ways: waters are generally colder in British Columbia than California, and a smaller human population means algae-producing nutrient pollution is less of an issue.

These factors all influence the way that sea otters interact with seagrass in BC waters. Often they target small "pocket" eelgrass meadows in nearshore areas and bays, rather than in estuarine habitats—and when they do, some of their chief targets are butter clams (*Saxidomus gigantea*) and littleneck clams (*Leukoma staminea*). As sea otters dig for them, they leave large holes in the sand and tear up the mat of roots created by the eelgrass.

This destructive foraging may have an upside. Researchers are exploring the possibility that the disturbance sea otters bring to eelgrass beds in British Columbia may have positive long-term effects, in the way that fires can ultimately be beneficial to forests.

Seagrasses in stable environments tend to reproduce by cloning themselves—that is, extending out their rhizome mat beneath the sand and growing new shoots from this expanding root mass. Much like stands of aspen trees that are actually one organism, a single seagrass plant can form a clonal meadow thousands of square meters in area (see page 100).

Crater excavated by a
sea otter seeking clams.

Coastal ecologists Erin Foster and Jane Watson are investigating their hypothesis that when otters forage for clams in eelgrass beds, root damage induces seagrasses to reproduce sexually by flowering—which they can do, just like the terrestrial plants that are their ancestral relatives—rather than only sending out clonal shoots.

Foster cites an ecological theory called the intermediate disturbance hypothesis, which suggests that plants in a stable environment do well to clone or reproduce asexually because "the genetic makeup of the parents is well suited to it, and therefore a genetically identical offspring should also be well suited."

However, if the environment is disturbed, says Foster, a plant's best strategy may be to respond by flowering and sending out pollen—a form of sexual reproduction that mixes its genes with those of other plants. The new genetically unique offspring potentially stand a better chance of being suited to the changing environment.

In other words, at seagrass sites that sea otters forage over for decades, the disturbance they bring could increase the genetic diversity of the seagrass. This is increasingly important because greater genetic diversity means greater potential resilience in the face of climate change–induced factors like warming waters and a related uptick in disease events (see page 71). It even appears that the pits left behind by foraging otters could open up space for new plant recruits to settle.

"Sea otters and seagrasses have coevolved," says Foster, "so it makes sense that seagrass would adapt in response to sea otter disturbance, and these disturbances may actually benefit the seagrass." ■

7 BEYOND THE BLOB
OCEANOGRAPHY AND OUR CHANGING SEAS

ON A ROUTINE trip to Calvert Island in the late spring of 2016, oceanography researcher Brian Hunt was startled by what he saw outside his seaplane window: an enormous milky cloud in Queen Charlotte Sound. "We could have been in the Mediterranean," he recalls. While a herring spawn event can have a similarly colorful effect on the ocean, his trained eye knew that wasn't the cause. Looking out over the swath of opaque turquoise, Hunt thought, as only a biological oceanographer might: "That looks like a coccolithophore bloom."

His hunch was right: researchers later confirmed that the event was a massive bloom of the single-cell, calcite-producing organisms. Similar sights were seen throughout the Strait of Georgia farther south in the summer of 2016, which appears to have been the first such bloom in the area in recorded history. Found throughout the world's oceans, coccolithophores are most abundant in subtropical zones (to see them up close, see illustration on page 45). The rare appearance of trillions of these tiny animals in British Columbia's coastal waters may be evidence of the shifting state of our oceans—and the event was all the more remarkable for being easy to spot, even from an airplane.

The birth of the Blob

Understanding the chemical, physical, and biological changes that under-
lie events like the one Hunt witnessed in midflight is not so simple. It
takes trained oceanographers, an array of specialized equipment (see
page 123), and a considerable amount of time spent crunching data.

Two years before, following an unusually calm and storm-free winter,
scientists had detected a large patch of warm water off the coast of Alaska.
A Washington State climate scientist had nicknamed the anomaly "the
Blob," and the name stuck. Like its Hollywood counterpart, it would grow
to become a disturbing and destructive phenomenon.

In the autumn of 2014, data from measuring buoys up and down the
western coast of North America had recorded a wide scattering of ele-
vated temperatures, all within the same period. Relative to the long-term
mean temperature, sea surface readings were 2.5°C higher on average
because of the Blob, with warming extending to a depth of 100 meters.
Ocean buoys off the Oregon coast recorded sea surface temperature
spikes of up to 7°C. "It was an extremely big change," recalls Hunt. "So
extreme, we wondered if there was some mistake in the data. We checked
and checked that it was the real thing."

Talk about these disturbing measurements dominated discussions at the annual State of the Pacific Ocean conference in 2015. Hunt presented Hakai Institute data collected from the waters around Calvert Island to tell the story of the unprecedented warm water mass at the northern extent of the record. "That data really helped us know when the Blob arrived on the coast of BC," says Hunt. "I think it was the first time it really came home to people how important it was to be able to collect data in the winter months, especially on the central or northern coast of British Columbia, which is hard to get to."

Known as a marine heat wave, the relatively stagnant, low-nutrient mass of water called the Blob affected an area greater than 9 million square kilometers—from the Bering Sea along the coast to Baja California. It contributed to an erratic and largely negative series of occurrences across this massive range: toxic algal blooms, altered spawning behavior in fish, mass strandings and starvation of marine mammals and seabirds, and a devastating outbreak of sea star wasting disease (see Chapter 4). Fishermen and recreational divers in California encountered many species accustomed to warmer waters, such as hammerhead sharks and ocean sunfish.

Jennifer Jackson is a physical oceanographer whose focus is the temperature, salinity, and oxygen levels of seawater on British Columbia's Central Coast. Her early research focused on the Arctic Ocean, where water temperatures have been on the rise since the 1990s. When she turned her attention to Pacific waters in 2015, she did so with an awareness that, as she puts it, "something weird was going on."

The Blob began when a formidable ridge of abnormally high atmospheric pressure—known to researchers with an alliterative bent as the Ridiculously Resilient Ridge, or the Triple R—parked itself over the northeast Pacific. Winds cool surface layers of the ocean similar to cooling a hot bowl of soup by blowing over it, and this high-pressure mass lowered wind speeds, resulting in a thermal buildup. The Triple R also

TOP: Oceanographic measuring buoy in the waters off Calvert Island.

BOTTOM: Sunfish (or mola mola), the world's largest bony fish, are sometimes seen in BC waters.

Stormy waters off Calvert Island's West Beach in October.

blocked seasonal winter storm activity for two years beginning in 2013, compounding the issue. Winter storms churn the layers of the ocean, and without such storms, the cold water below 100 meters deep stays put and doesn't help cool waters nearer to the surface.

"When storms don't happen, the heat that's accumulated through the summer is essentially trapped," says Jackson. "And that's what we think caused the Blob: really abnormal winter periods with few storms."

A salmon mystery in Rivers Inlet

Oceanography is a richly interdisciplinary field. Among other things, oceanographers try to make sense of the physical and chemical properties of seawater, the life within it, and its connection to the rest of the planet. But understanding the causes and effects of atypical events like the Blob is difficult, because the Pacific Ocean—the world's largest and deepest ocean—is a complex beast even on an ordinary day.

Anomalies need to be teased out from overlapping events such as the irregular appearance of El Niño, which increases sea surface temperatures across the Pacific and can shift weather patterns from Africa to Nova Scotia, and the Pacific Decadal Oscillation, which alters ocean temperatures in the North Pacific. Then there are other factors affecting oceanographic conditions, such as the California Current, which brings cooler water south from British Columbia to Baja California, and the consistent phenomenon of upwelling in coastal areas—where offshore winds push warm surface waters away from the land, drawing up colder, nutrient-rich water from below.

Physical oceanographers keep their eye on patterns like these, attempting to untangle them from yet other phenomena—many linked to global warming—that can cause events like the Blob. Through the Hakai Institute, Jackson's own oceanographic research is focused on Rivers Inlet, a narrow, 45-kilometer-long glacial fjord on the Central Coast. Her work there indicates that the impacts of the Blob may be more long-lasting than many researchers had previously suspected.

Rivers Inlet is surrounded by temperate rainforest in the territory of the Wuikinuxv Nation. At one time, Rivers Inlet and nearby Smith Inlet together made up the second-largest sockeye salmon run in British Columbia and supported a thriving fishery. But their numbers began to decline in the 1970s and ultimately collapsed in the early 1990s: salmon that used to return in the millions were reduced to a few thousand. Neither the impacts of commercial fishing or logging in the watershed appeared to explain their slide into near-extinction, which had a devastating effect on the Rivers Inlet ecosystem—particularly on salmon predators such as grizzly bears, eagles, and wolverines, and the Wuikinuxv people themselves. During the late 1990s, the collapse forced the Wuikinuxv people to purchase food from Port Hardy, 130 kilometers away on Vancouver Island.

Despite a ban on commercial fishing that began in 1996, salmon populations have never returned to historic levels, and the primary cause of the Rivers Inlet salmon collapse has remained obscure. Now, coupling 70 years of oceanographic data with recent measurements she and other researchers have gathered, Jackson believes they may have discovered part of the answer: abnormally warm ocean temperatures.

In the summer, northwesterly winds drive coastal waters offshore and allow deeper ocean waters to come into inlets and other coastal areas, a phenomenon known as upwelling. Jackson believes that the deeper layer of warm water upwelling into Rivers Inlet during the years of the salmon collapse could have caused poor feeding conditions—zooplankton on which salmon feed are less fatty and nutritious in warmer waters—and resulted in low survival rates.

"We looked at previous periods when there's been abnormally warm water, and those are linked to times when the sockeye salmon collapsed in Rivers Inlet," says Jackson. "It's speculative, but we found a really strong correlation with what happens offshore and what we see in the deep water of the inlet."

Jackson and Hunt, along with many other researchers, are working to understand how a potential long-term shift to warmer waters will impact creatures throughout the coastal marine food web. Pacific salmon, not surprisingly, are one indicator of special interest. How they respond to these changing conditions is of concern not only to scientists but to Indigenous groups, fishery managers, and coastal residents from Alaska to California.

"People get anxious and worried about salmon, and rightly so," says Hunt, pointing out that most salmon populations have been in various states of decline for decades. "Warm conditions are generally not great for salmon. That's because they have a temperature threshold above which they don't do well, and because of how these new conditions affect

Salmon are an important part of the diet of coastal gray wolves and eagles, among many other species.

the food web, which means less growth, less energy for spawning, less production for salmon."

Salmon are an illuminating species to focus on when studying the effects of climate change on marine ecosystems because of the relative abundance of information on them. Hakai Institute researchers have taken an integrated approach to studying salmon in British Columbia, amassing data that will help to understand their health through the entire migration cycle. "With our instrumentation, we're able to get to the bottom of the food chain and really link that in a dynamic way to salmon health," says Hunt. "That's been done here and there elsewhere in a spotty fashion, but never consistently as we've been able to do in recent years."

He and his colleagues on Calvert Island are currently working to make the most of field measurements of oceanographic data like temperature, salinity, and chemistry by coupling them with samples of plankton and marine microbes. "This allows us to make correlations between environment and biology, which is often very messy," he says. These correlations are beginning to reveal some of the fascinating dynamics of marine food webs (see Chapter 3).

A juvenile coho salmon off Lelu Island in northern British Columbia.

Hunt's current work involves putting animals such as salmon and plankton from the wild into controlled environments in the lab for experimentation. "We replicate ocean chemistry in the lab and test animal responses to changes in temperature, diet, and acidification. This allows us to begin to narrow down the key things that are affecting them."

While he doesn't expect salmon to become extinct, Hunt believes some populations that are currently not faring well on the west coast may disappear from places where we're used to seeing them. Meanwhile, species that live in colder water, such as the Bering Sea, appear to be thriving, he says. "We're seeing a geographical shift in the production of salmon."

Beyond the Blob

Marine heat waves are a global concern. Scorching temperatures and wildfires in Australia made international headlines beginning in late 2019, but less well known are the ocean heat waves that have struck the waters off the southeast Australian coast since 2015. The Tasman Sea is warming at four times the global average, and 95 percent of the kelp forests (see Chapter 5) on the east coast of Tasmania have been wiped out by rising temperatures.

In the northeast Pacific, the changes researchers have seen following the advent of the Blob continue to reverberate through the ecosystem.

The Limpet, an underwater observatory off Quadra Island that offers researchers a wide variety of oceanographic data.

As Jackson has analyzed the layers of water in Rivers Inlet—surface (5 to 10 meters deep), intermediate (10 to 140 meters), and deep (140 to 330 meters)—she's come to suspect that warm water conditions will continue for some time in the wake of the Blob.

"Once I separated Rivers Inlet into layers and started tracking heat, I realized that in that deep water, there was this huge pulse of heat that came in with those northwesterly winds," says Jackson. "A lot of heat from the Blob that was out in the open ocean was stored at depth, and every year we get upwelling, so the inlet still really hasn't cooled down."

Deep water in the region continues to measure half a degree warmer than the long-term average, which—thanks to the overlapping impacts of El Niño and recurring marine heat waves—may be part of a new normal for the Central Coast. "I don't really see a pathway for the waters here in Rivers Inlet and other areas of the coast to cool down unless the source water in the Pacific changes."

This may not mean that coastal waters from British Columbia to California are going to turn turquoise every year due to exotic plankton blooms—but unusual oceanographic conditions are going to present serious and ongoing challenges for a wide range of marine and terrestrial species, from the very large to the very small.

Researchers from Hakai and Simon Fraser University sampling juvenile salmon in Johnstone Strait.

TOP: Robotic glider being deployed by a research team.

BOTTOM: Katie Pocock dropping a CTD monitor into the water column in Baynes Sound, south of Quadra Island.

THE SCIENCE OF SAMPLING

Since the 1930s, the University of British Columbia and Fisheries and Oceans Canada have collected water samples off the BC coast by submerging collection bottles—now known as Niskin bottles (see page 50)—on a line. In the 1970s, new sensor technologies helped refine data collection by adding an instrument called a CTD (for conductivity, temperature, and depth) to these water sample tests.

Over 30 countries participate in the global Argo Program, which since 2000 has become the world's major source of deep ocean data on temperature and salinity. Argo has enabled this collection of robust oceanographic data via autonomous robots, called profiling floats. Over 3,800 Argo floats—at a cost of about $40,000 each when deployment and operational costs are factored in—descend to depths of up to 2,000 meters, collecting data on ocean temperature and salinity as they rise to the surface, where they transmit data to researchers via satellites. These autonomous floats make their oceanographic data—including information crucial to weather forecasting and predicting overarching climate events such as El Niño—available to researchers within 24 hours.

A new program is taking marine sampling to the next level via a fleet of robotic gliders—autonomous winged robots that "fly" underwater from the shallows to the depths and back again. Developed by the Canadian-Pacific Robotic Ocean Observing Facility (C-PROOF), the project hopes to gather more data on what its researchers call "poorly understood" qualities of the ocean, using gliders to collect data on marine life, ocean turbulence, and nutrient levels

Ocean acidification research at Hakai's Quadra Island Ecological Observatory.

as they travel up to 30 kilometers per day. A broad collaboration that includes researchers and technicians from multiple institutions, the long-term goal is to create continuously sampled lines in BC waters; one of these will run west from Calvert Island about 150 kilometers to the shelf break, where the North American continental shelf slopes away to the open ocean.

Along with better climate and weather predictions, researchers hope data from C-PROOF gliders will help experts make more accurate assessments of fish stocks and move "remote coastal communities and the country as a whole" toward carbon-free energy supplied by ocean wind, tides, and waves. ■

THE ACIDIFYING OCEAN

In the early 2000s, many thought that having the ocean serve as a repository of human-generated carbon was a good thing. "With the exception of a few marine chemists, generally the idea back then was that the ocean was helping reduce the greenhouse warming effect," says Wiley Evans, a chemical oceanographer and Hakai Institute researcher who studies carbon dioxide (CO_2) levels in seawater. "Perhaps carbon dioxide absorption into the ocean reduced some immediate atmospheric warming effects, but we've since learned that's not the whole picture."

The ocean absorbs up to a third of the industrial emissions of CO_2 that humans create via the burning of fossil fuels. One recent study showed that between 1994 and 2007 alone, the ocean absorbed 34 gigatonnes of CO_2. According to one of the study's authors, Richard Feely of the US federal National Oceanic and Atmospheric Administration

LEFT: The well-known Pacific oyster, actually a non-native species originally from Asia. Researchers are attempting to understand the impacts of acidifying oceans on oysters and many other shellfish species.

RIGHT: Top, a weathervane scallop. Bottom, a close-up of the eyes of a different scallop species, likely a rock scallop or a swimming scallop.

(NOAA), on average this is equivalent to the weight of 2.6 billion Volkswagen Beetle cars being dumped into the ocean each year.

Prior to the beginning of the industrial era in the mid-18th century, there were natural swings in atmospheric CO_2 content. Those earlier swings are sometimes incorrectly cited to prove that humans are not responsible for the industrial-era increase in atmospheric CO_2, nor for the global warming that has occurred as a result.

This is not the case. Ice cores taken from Antarctica, Greenland, and elsewhere show that in the preindustrial era, CO_2 levels ranged between 200 and 300 parts per million (ppm). Current data from NOAA's monitoring program at Mauna Loa Observatory in Hawaii show that industrial-era emissions have pushed us to about 412 ppm. "This is nearly one hundred and fifty ppm over anything seen in the past eight hundred thousand years," says Evans.

As all that CO_2 enters the ocean, it forms a weak acid that ultimately lowers the pH of seawater. Scientists say that this process transforming our oceans is happening faster than any change in marine CO_2 chemistry in the last 50 million years, with a range of potential consequences for marine life.

Some organisms, such as certain varieties of phytoplankton, may do slightly better in these emerging conditions. For many others, a more acidic environment will potentially impact their ability to feed and reproduce, or threaten their very survival. Research by the University of Washington and NOAA on Pacific salmon indicates that acidic seawater disrupts their highly developed sense of smell, which they use to detect food, avoid predators, and navigate to their natal streams. Meanwhile, sea creatures with carbonate shells or skeletons—such as oysters, mussels, and corals—have difficulty producing shells in corrosive seawater.

"There are really clear examples of negative consequences for certain organisms, such as oysters and other bivalves," says Evans. "There is experimental work and observational information that shows that these organisms are going to have a very hard time in a future corrosive seawater environment."

Findings like these have understandably generated considerable anxiety among shellfish growers and harvesters. Understanding the way that acidification will affect sea life, however, is a complicated affair. "We're intensively trying to understand the negative consequences for bivalves and other organisms," says Evans. "We do know there's a big gradient across which these impacts are going to take place, and how they unfold."

The ultimate solution for all marine species, says Evans, is to drastically reduce CO_2 emissions. In the meantime, he hopes that researchers will be able to give "headlights" to shellfish growers and other harvesters so they have some notice of coming conditions in an increasingly acidic ocean. ■

Scientist Helen Gurney-Smith and her team gathering shellfish for testing from a research raft off Quadra Island.

8 LICENSE TO LOOK

THE JOY OF BIOBLITZES

ASK A CHILD to imagine what ocean scientists do, and they might come up with a fairly accurate sketch of a marine bioblitz: researchers poking around tide pools in rubber boots, drawing up buckets of critter-filled seawater; scuba divers plucking crabs, kelp, and sea cucumbers off the seafloor; lab-coated biologists spending long hours peering into microscopes. For the more than two dozen scientists who participated in the first Hakai-MarineGEO bioblitz in the summer of 2017, scenes like these were common—and often accompanied by kids-at-the-beach giddiness as they encountered species never before recorded by science.

"That blitz reinvigorated a passion I had early on when I got into the field," says Matt Whalen, a community ecologist and Hakai Institute postdoctoral fellow who helped manage the event. "It's that excitement you get thinking, 'What are all these amazing creatures in the ocean and how many of them are there?'"

Whalen calls the Central Coast a "cold water hot spot" of biodiversity, due to its combination of glacial activity, kelp forest ecosystems (see Chapter 5), and a complex coastline of islands, bays, and fjords. Along with marine mammals like sea otters and whales, the area retains an impressive and often dazzling diversity of invertebrates, seaweed, and fish. "Part of the amazing thing about that bioblitz," says Whalen, "was being able to bring in biologists and taxonomists who know a particular group of organisms really well, and basically set them loose into this new place that they've never been."

Gustav Paulay, who helped to coordinate the bioblitz's activities and infrastructure, was one of these experts. He calls the area around Calvert Island "a biologist's dream." An invertebrate zoologist who specializes in biodiversity surveys and a curator at the Florida Museum of Natural History, Paulay is no stranger to the Pacific coast: he got his PhD in zoology at the University of Washington. "When I was a student there, we thought we knew the northeast Pacific marine biotas very well," says Paulay. "Surveys like the one at Hakai show that we know little."

An urban origin

Part scavenger hunt, part science marathon, bioblitzes are more formally described as large-scale biodiversity surveys. Researchers and citizen scientists at a bioblitz aim to record biological life at a particular place and time using a variety of formats: photographs, specimens, and DNA sequencing. The term "bioblitz" came from the first biodiversity survey marathon held at Kenilworth Park and Aquatic Gardens in Washington, DC, in 1996, where participants collected and identified a thousand species in only 24 hours. Sam Droege, a wildlife biologist and one of the originators of the bioblitz, recalled the inspiration for holding that first event in an interview with *National Geographic*:

> Our BioBlitz idea was that it should be an event. We would attract biologists with coffee, which is all it takes, and a whole bunch of other interesting "'ologists" as well. The key was that they weren't restrained, they were allowed to move back to their primitive state, the ancestral biologists who actually went out and tried to find things, instead of filling out forms, or gridding on a plot, or counting the number of tarsal segments and measuring them to within 15 microns.
>
> The idea: do what you do naturally—which of course is why they became biologists. Go out and find critters, or a plant, or a fungi, or an isopod, whatever it is that got you to be a biologist in the first place.

Droege credits the invention of the name "bioblitz" to Susan Rudy, who at the time of the first event was a Kenilworth naturalist. Since then, research institutes and conservation groups around the world have used the name for time-limited surveys that set regional baselines

TOP: Nearshore researchers surveying fish caught in the shallows.

CENTER: Jennifer Walkus sorting species for a Hakai bioblitz.

BOTTOM: Gillian Sadlier-Brown with a sample of sea snails.

for biodiversity. These detailed catalogues of an area's vital signs help researchers better understand the effects of environmental changes, whether from habitat shifts, natural events, or human-caused pollution.

Many bioblitzes last only a single day, but the 2017 survey, hosted at the Hakai Institute's Calvert Island Ecological Observatory, was three weeks long. It was an ambitious project made possible not only by participating scientists but by a large crew of technicians, support staff, and volunteers from local First Nations groups, among others. The event was part of the Smithsonian Institution's equally ambitious Marine Global Earth Observatory (MarineGEO) BioBlitz program. MarineGEO collaborates with researchers around the world to document biodiversity in coastal areas—the areas of the planet where human populations are often at their densest—and to understand the ways in which nearshore marine life is changing.

From the Lab Queen's microscope to eDNA

Veteran marine biologists and taxonomists used their time at the Calvert Island observatory to examine samples from 255 different sites in and around its coastal waters, identifying several thousand specimens representing approximately a thousand species. This is an impressive accomplishment given that the total number of marine species currently known to science—from microscopic protists to algae to worms to whales—is about 250,000.

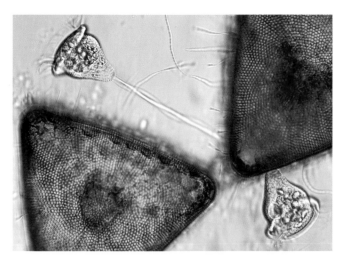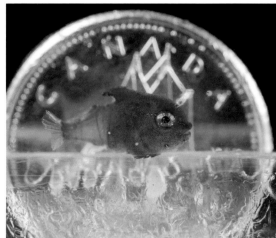

LEFT: Ciliates on diatoms.

RIGHT: A baby spiny lumpsucker.

Whalen, who was tasked with reporting on the final results of the event, says each day of the blitz brought gasps of astonishment and whoops of joy from taxonomists who spotted species they'd never seen before—or not seen in such abundance.

Leslie Harris of the Natural History Museum of Los Angeles County—the 2017 blitz's exuberant, unofficial "Lab Queen"—clocked up to 18 hours a day in front of a microscope, identifying more than 300 polychaete worm species. "Some people see me as a lab slave and that I have this invisible chain and manacle tying me to the microscope," she told a *Hakai Magazine* interviewer, "but I prefer to think of myself as the Lab Queen with my minions bringing me samples."

Standout specimens for Harris included colorful new species like an orange *Eusyllis* worm with enormous eyes that she called Joe Cool (after Snoopy's alter ego), and the Desi Arnaz worm (of the family Dorvilleidae), so nicknamed by Harris for its resemblance to the frilly sleeves on the Cuban shirts worn by Arnaz's band. At least 40 of the worms she catalogued had either never before been recorded for the area or were undescribed.

University of Washington amphipod expert Craig Staude, meanwhile, was captivated by his first encounter with a live *Erichthonius brasiliensis*, a red-eyed species he'd seen numerous times before, but only in preserved, bleached-out form. "Specimens in gray sandy tubes have never seemed impressive," Staude wrote of his experience. "Here at Hakai however I had the chance to see this amphipod several times, living in its tube or

Top left, the frilly Desi Arnaz worm (genus *Iphitime*), as coined by Leslie Harris. Top right, a worm (genus *Cheilonereis*) that shares shells with large hermit crab species. Bottom left, a fluorescent-green, free-swimming paddle-worm. Bottom right, a parasitic isopod in the genus *Rocinela*.

actively crawling around. I was fascinated by its red eyes and powerful stance when protecting its tube."

Estimating the actual number of species in the ocean is notoriously difficult and a matter of ongoing debate. One estimate is that there are 700,000 to 1 million multicellular species in the ocean—setting aside microbial organisms like viruses and bacteria—leaving between one-third and two-thirds of them yet to be described. Paulay believes the number of multicellular marine species to be at least 1 million, with three-quarters of them currently unrecorded. In either case, there remains a lot of work yet to be done in understanding the complexity of ocean life.

"We've spent a lot of time and effort exploring the cosmos and I would say there's still a lot—a lot—more to be understood about our planet," says Whalen. "Even in shallow water just right outside our door, there are new species to be discovered."

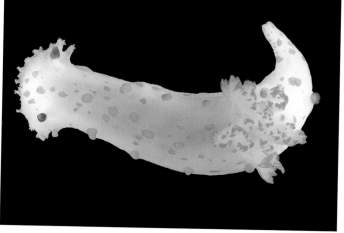

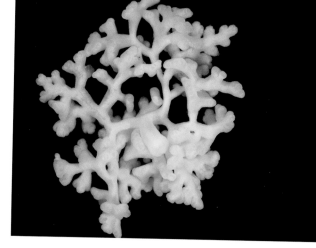

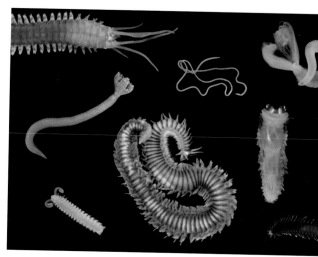

The vast number of species yet to be identified is complicated by the fact that some organisms already known to science are categorized incorrectly. Proper identification is difficult when many species are "cryptic"—meaning close to identical in their form, also known as morphology, and nearly impossible to distinguish by sight. (One example is the *Henricia* genus of sea stars, a graceful thin-limbed starfish whose species are often confused.) Early European researchers were also guilty of frequently, and inaccurately, lumping Pacific varieties in with species from the Atlantic; these now "misknown" species need to be corrected.

Innovations in genomics technology have made this work easier, enabling scientists to quickly identify and compare DNA of the organisms they find in the field. Large numbers of them can even be identified at the same time by analyzing biological material in seawater or sediment, an emerging technique known as environmental DNA or eDNA.

Top left, a clown nudibranch. Top right, a parasitic barnacle in the genus *Dendrogaster* that makes its home in a leather star. Bottom left, a juvenile heart crab. Bottom right, a collage of segmented worms.

"All predators, fish, and bugs and everything are shedding little bits of skin particles and DNA into the water," explains Matt Lemay, a senior research associate at the Hakai Institute whose work uses genomics to study biodiversity. "We do a lot of work with eDNA now, which means that we can scoop up a liter of seawater and then sequence DNA from that."

Many institutions are unable to do this genomic work because of the time, expense, and expertise required. The Hakai-MarineGEO bioblitzes take care of the need for expertise by bringing in some of the world's top taxonomists that specialize in particular organisms. "We end up with this really great thing where we have the world's best taxonomic experts telling us what something is," says Lemay. "From every specimen that we collect, we get a photograph as well as this large, standardized DNA sequence that can be used to identify that species in the future. Then the specimen itself goes into a museum for the future."

These samples become part of a biological record of an area—which is then available to researchers and taxonomists, Lemay says, "forever." Lemay is helping the Hakai Institute curate a genetic database for all of the life on the Central Coast, which in turn is part of a global data-set shared with researchers and MarineGEO sites around the world. Bioblitzes feed directly into that and offer a surprising amount of new information: about one-fifth of the specimens that the Hakai Institute samples, says Lemay, haven't had their DNA sequenced before. "These

Samples of genetic codes showing DNA bases next to two different crab species. Researchers use this kind of information to make accurate species identifications.

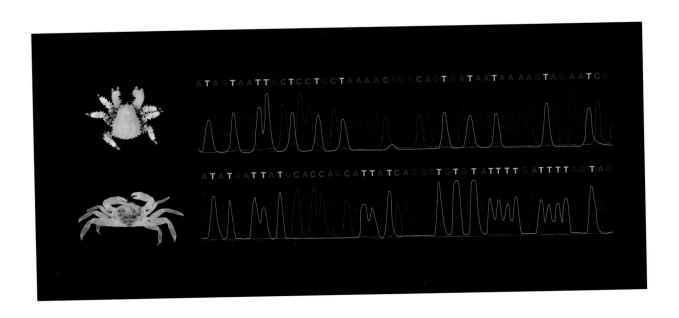

are completely new records that we're adding to this global database," says Lemay.

Once an organism's genetic information is recorded, that sequence becomes a reference tool for further work, allowing researchers to identify many of the animals that live in that area. "Then if you ever have an unknown organism, like a piece of tissue from a stomach or something, and you don't know what it is," says Lemay, "you can sequence it and try to match it to the database to figure it out. This provides a really powerful tool for monitoring biodiversity and ecosystem changes."

Since the inaugural 2017 event, subsequent bioblitzes at Hakai have narrowed their focus. One looked at seagrasses; one focused on terrestrial organisms (which sampled more than 4,000 insects); and another explored biodiversity in deep water using scuba divers and a remote-operated vehicle.

LEFT: Researcher Sung Jo showing off a jar of iridescent ground beetles.

RIGHT: Bees and other insects collected during a terrestrial bioblitz.

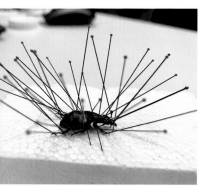

TOP: Collecting carrion beetles for a bioblitz.

BOTTOM: Prepping a carrion beetle for display.

Saving pages from a burning library

Perhaps the most pressing reason for bioblitzes is the rapid rate at which species—aquatic and terrestrial—are currently declining or even disappearing altogether. Like many other researchers, this has fueled Gustav Paulay with a passion for taxonomic and genetic identification through bioblitzes.

"We live at a strategic time," he says. "Humanity is messing with the planet on an unprecedented scale, causing an enormous amount of change. Our greatest impact by far will be the loss of biodiversity and associated loss of habitat and ecosystems."

This loss of biodiversity is irreplaceable, says Paulay, noting that we are losing "millions to hundreds of millions of years of accumulated evolutionary innovation. Think of it as the greatest of all libraries burning down."

For Paulay, "the greatest meaningful action" that biologists can have is to archive as much information in the global library of biodiversity as possible. He laments that few young people choose careers in natural history or taxonomy, which despite using some of the most advanced genomics technology available, seem to many students like outdated, "nineteenth-century" fields. "Just at the time when it is most needed, few folks are out there collecting the data needed by future generations—making recordings, taking samples, preserving the genomes of our dinosaurs and dodos."

In this context, the celebratory nature of a bioblitz punctuated by whoops of joy marks a rare chapter in the story of a changing planet. It's a welcome pause that, says Whalen, allows scientists and citizens to be amazed and moved, and to remember what we're working to protect. "Events like this give us a sense of 'ocean optimism,' which is a term coined in the past few years by researchers and marine conservation advocates," says Whalen. "We hear so many stories about the doom and gloom, and that is, of course, very much real and worthwhile—but we also have a lot to be grateful for."

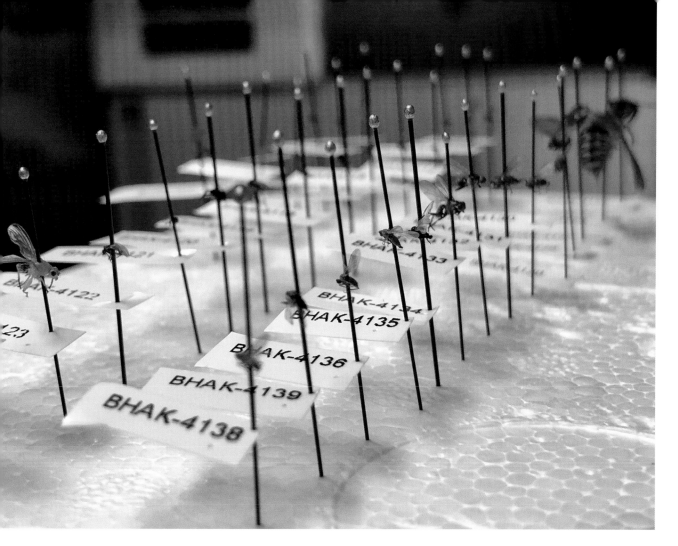

TOP: Small flies and a wasp pinned after being sampled for DNA analysis.

BOTTOM: Fishing spiders, like this one found in a bioblitz, are one of Canada's largest spider species. In freshwater ecosystems, they float on the surface and can dive to catch fish and tadpoles.

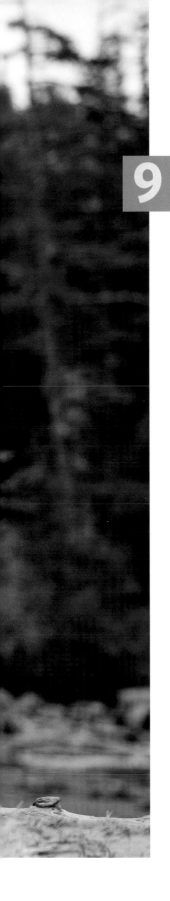

9 ON THE SHORE

MARINE SUBSIDIES AND DUNES

BRIAN STARZOMSKI WAS on his hands and knees counting plants in an island forest when he looked up and saw a wolf watching him. Starzomski is a community ecologist—a researcher focused on the interactions between species—and he was in the midst of gathering data in the Goose Islands on British Columbia's Central Coast. Despite the archipelago's distance from the mainland, says Starzomski, it is "crawling" with gray wolves (*Canis lupus*).

Around 2007, the enterprising predators managed to swim their way over to the islands to feast on black-tailed deer (*Odocoileus hemionus*). The deer arrived on the islands about 20 years ago and radically altered the vegetation, transforming nearly impenetrable, bushy thickets into mossy woodlands. With the arrival of the wolves, black-tailed deer have all but disappeared from the Goose Islands. The vegetation is making a come-back, but while wolves will occasionally eat berries, they are carnivores: plants alone aren't enough for them to survive. Given the disappearance of what was once their primary food source, one might expect a Goose Islands wolf to be deeply interested in a mammalian visitor—especially a relatively clawless and fangless one.

In his telling of the encounter, however, Starzomski describes the wolf as simply "curious." It seemed its dietary needs were being met, somehow. In fact, the wolf's way of life marked it as one of the agents

PREVIOUS SPREAD: Coastal gray wolf keeping a watchful eye on a beach.

ABOVE: The remains of marine organisms brought on land by river otters and other creatures can turn up in incongruous places. Left, urchin tests on a mossy knoll. Right, abalone shells on the forest floor.

and beneficiaries of the very phenomenon he had come to the islands to study: namely, marine subsidies.

The 100 Islands project

A marine subsidy refers to the flow of nutrients from sea to land. There are many ways this transfer can happen. Seabirds leaving guano on land and humans leaving shell middens are examples of ocean nutrients altering and enhancing terrestrial landscapes. Nitrogen and other nutrients in sea mists can precipitate out over land and "subsidize" coastal soils.

The flow of nutrients goes the other way as well, of course: glaciers, rivers, and forests all contribute nutrients to coastal waters (see Chapter 2). Both types of nutrient flow, and their effects on biodiversity, are of interest to scientists. As the economic shadings of the term suggest, marine subsidies can promote productivity in areas that would remain underdeveloped or barren without support from the ocean—particularly in desert-biome islands such as those in Mexico's Sea of Cortez. But marine subsidies can also have a significant impact on areas already rich in resources. One of the more well-known examples is the phenomenon of the "salmon forest," where trees can be well fed by the carcasses of spawning salmon dragged up onto the land by bears and wolves. Sitka spruce near some salmon-bearing streams in Alaska grow up to three times faster than those that are not—reaching a diameter of 50 centimeters in 86 years rather than 300 years.

Less studied is the phenomenon of marine subsidies on the remote islands of the Central Coast where, unfortunately for resident wolves, there are few salmon. Starzomski, the director of the School of Environmental Studies at the University of Victoria, was keen to see more research done there. "We were interested in studying the effects of marine nutrients on these coastal islands where there really aren't very many salmon, or no salmon at all," he says.

The origins of the project can be traced to Starzomski's colleague John Reynolds; he was equally keen to explore the islands off the Central Coast, which are notoriously difficult to reach. Reynolds is a professor of aquatic ecology at Simon Fraser University and has served as chair of the Committee on the Status of Endangered Wildlife in Canada. In 2011 he partnered with the Hakai Institute to study the ways that island biogeography—the richness, distribution, and colonization of islands by various species—linked with marine subsidies in the Goose Islands. His

Hungry as a wolf: coastal gray wolves will dig for clams, eat barnacles, and pluck squid from the water at low tide.

biodiversity research there grew into an ambitious effort to study the way that nutrients from the sea influence insect, bird, mammal, amphibian, and plant populations on and around a hundred of the Central Coast's outlying islands.

With collaboration from the University of Victoria and the Hakai Institute, Reynolds assembled other principal investigators—including Starzomski, conservation scientist Chris Darimont, and geographer Trisalyn Nelson—and a dozen graduate students to investigate the links for a three-year study. The project launched in 2015. Because of the islands' remote location, logistical support was essential—and the Hakai Institute's Calvert Island Ecological Observatory was well placed and willing to supply it.

Sara Wickham, a member of Starzomski's team, can attest to the challenges. "Planning a three-month-long camping trip for 15 to 20 people is no easy task," she wrote in an article on the project. "Logistics such as remote campsite space, food, camping equipment, science equipment, transportation, and toilets are just the start. This may be why the project earned the nickname '100 issues' around Hakai Institute headquarters."

Wickham was also well positioned to do that work. At the time, she was a master's student interested in the overlaps between marine and terrestrial ecology.

"We tend to think of marine and terrestrial environments as having wholly distinct ecosystems," she says, "but there is actually a great deal of movement between the two."

LEFT: The 100 Islands project explored the ways that nutrients from the sea influence insect, bird, mammal, amphibian, and plant populations on Central Coast islands and islets.

ABOVE: Sea wrack is one of the many ways that coastal terrestrial ecosystems receive nutrients from the sea.

"Wracking" up nutrients

One avenue for exploring these overlaps is the food web. Food webs map the complex connections and predator-prey relationships that make up the flow of nutrients in an ecosystem. On islands in the Central Coast, one component of the food web is seaweed or kelp. Kelp is a significant source of omega-3 fatty acids, an essential nutrient for many organisms.

Bull kelp washed up on Seventh Beach on Calvert Island.

Kelp forests (see Chapter 5) off the Central Coast are primarily made up of two types of kelp: giant kelp (*Macrocystis pyrifera*) and bull kelp (*Nereocystis luetkeana*). When these two seaweeds reach the end of their respective life cycles, they die off and float away. Storms can also "rough up" and tear off portions of these plant-like organisms, or pull them completely off their seafloor holdfasts while still alive. Along with torn or uprooted eelgrass (*Zostera marina*), a marine plant distinct from kelp, this material eventually gets deposited on beaches by tides and storms. Once there, it is known as wrack or sea wrack.

Exploring the effect that wrack has on island food webs in the Central Coast was Wickham's job. After a boat trip from Port Hardy on northern Vancouver Island to the Calvert Island Ecological Observatory—a trip that, depending on weather, can be a two- to eight-hour boat ride—Wickham and her colleagues headed out to islands within a two-hour range from there. After setting up a central camp, up to 14 researchers would use a fleet of Zodiacs to zip out to outlying islands to gather data on birds, mammals, insects, and plants.

In Wickham's case, that meant weighing a lot of seaweed—a lot of it "pretty rotten"—and categorizing it, if it was still recognizable. "We just did that over and over for two weeks," says Wickham, "and then we'd head back to Calvert Island for two or three days to have hot showers, get food, do laundry, and check the Internet. Then we'd head out for another two weeks."

The 100 Islands team repeated their two-week data-gathering forays for three months at a time. The relentless fieldwork was not without its upside. Back at their base camp, researchers would share food around nightly beach fires and talk about what they'd seen. "This was a great way to share information," recalls Wickham. "Usually when you go into the field, you're with one or two research assistants and you're all focused on the same thing. In this case we were all looking at different things, and that made for lots of interesting campfire talks. One person would share something that they'd seen in passing because it was a neat or odd observation—and it would spark a great conversation, or even turn out to be vitally important to someone else's research."

Sara Wickham, left, and Beatrice Proudfoot using a quadrat to track sea wrack on Calvert Island's North Beach.

Feeders on sea wrack, amphipods like these—also known as sand fleas or beach hoppers—in turn can provide a nutritious snack for coastal wolves and bears.

While Wickham's particular area of marine subsidy research may have been smelly and exhausting work for her, the sea wrack she studied is a paradise for organisms nearer the base of the food web. These are the invertebrates who call the sea wrack their home, including flies and beetles. The "top guy" among these invertebrates, says Wickham, is the amphipod. "It's those beach fleas that you see, the sand hoppers." Colloquially known as sand fleas, amphipods aren't insects at all, but tiny crustaceans: "little terrestrial shrimps, basically."

Wickham adds that amphipods aren't too fussy about their food. They'll eat any seaweed provided it's "a little bit old. They like their seaweed to be degraded for a week." Once sufficiently softened, the amphipods gobble it up, absorbing in the process the fatty acids—which make the crunchy little crustaceans into a nutritious meal for the animals that eat them in turn. These comprise a colorful array of critters. Songbirds, shrews, and mice all eat amphipods, as do a number of insect species.

Christopher Ernst, a community ecologist and entomologist who worked on the project with John Reynolds, says the "flashiest" of these insects is the pictured rove beetle (*Thinopinus pictus*). Over two centimeters long as an adult, it is pale yellow with black markings. "They hide under the sand, above the tide line, during the day and emerge under cover of darkness to feast on the hoppers," says Ernst. While amphipods seem unable to catch a break from terrestrial predators day or night, their bad luck was a boon to 100 Islands researchers—proving an important link between marine subsidies and island food webs.

TOP: University of Victoria and Hakai Institute researchers sorting through sea wrack on Goose Island.

BOTTOM: Katie Davidson inspecting a small mammal trap on Calvert Island.

Amphipods are even part of the diet of island wolves, like the gray wolf that was sizing up Starzomski in the woods. Compared to their counterparts on the mainland—who often compete with grizzly bears over salmon—gray wolves of the outer coastal islands get more of their food from marine sources, up to 85 percent. Adaptation to their ocean-bound habitat has even made them genetically distinct from mainland wolves. Along with hunting for mink and river otters, wolves will dig for clams, eat barnacles off logs, pluck squid from the water at low tide, and scavenge dead seals and even whales that end up on the beach. They will also make a meal of those jumpy little amphipods, which are most active and abundant in the summer. Wolves are not the only large mammals that will do this: hungry coastal bears will also snack on amphipods.

"You get a big pile of kelp that washes up that weighs several tonnes, it's all wrapped together and it has tens of thousands, hundreds of thousands of these little beach hoppers on it," says Starzomski. "And when you have that number of little tiny things a little less than a centimeter in length, that can make up a big meal."

Marine subsidies and resilience

The project discovered other examples of the effects of marine subsidies. Ernst found that sand beneath the wrack holds more moisture and nutrients, creating favorable habitat for plants like beach pea (*Lathyrus japonicus*) and dunegrass (*Elymus mollis*); these plants germinate faster and grow larger on beaches with large wrack deposits. Researcher Debora Obrist, a PhD candidate working with Reynolds, found that songbird populations are higher on Central Coast islands with more marine subsidies—specifically, on islands more laden with sea wrack.

Scientists working on the 100 Islands project also found evidence that river otters contribute significant subsidies to the island landscape by foraging for fish and marine invertebrates in the sea and leaving the remains of their prey, and their own scat, on land.

"River otters bring in huge amounts of marine subsidies, and we can see evidence of this by looking at changes in soil and plant nutrients along a gradient of 'ottery-ness,'" says Ernst. The most "ottery" places are latrine sites on shorelines, which have extremely high levels of marine-derived nitrogen. Plants that grow on and around these zones are very nutrient-enriched, says Ernst—but they also are stunted and lack

TOP: This deceased juvenile humpback whale on Calvert Island provides an outsize infusion of nutrients to the local ecosystem, feeding everything from wolves to amphipods.

BOTTOM: A close-up reveals a species of acorn barnacle that specializes in making its home on whales.

diversity. "The dominant types of vegetation near otter latrines are moss and algae, and there tends to be a lot of bare soil."

River otters also "make a mess" of shoreline habitats, says Ernst: they dig, scratch, burrow, roll, trample, and rip out plants. They even sweep away leaf litter, often leaving it, he says, "in tidy little piles." The chaos they create is off-putting to many ground-dwelling insects and spiders; the only ones that don't seem to mind are slugs. Looking up to the trees, eagles are another factor. Eagles will use the same nests for generations, and over time the terrestrial areas below them are likely to be significantly altered by their presence.

Exactly how these inputs by otters, eagles, wolves, and sea wrack affect the biotic communities on these islands is still a matter of ongoing examination by researchers—but as with the salmon forest, there is a clear link between healthy marine ecosystems and the fortunes of plants and animals on land. And while nutrient links can boost the resilience of ecosystems in the face of change, Wickham cautions that the blurring of boundaries between sea and land also means vulnerability for both.

"It's important to recognize that environmental disasters like oil spills will affect marine and terrestrial ecosystems through these land-to-sea connections," says Wickham. "When we plan for conservation management goals like marine protected areas, we really have to recognize that they won't work if we're not considering what's happening on the land, and vice versa."

While she laughs at the humbleness of studying rotting seaweed, Wickham is happy to be contributing to an ecological understanding with deep roots in coastal British Columbia. "This knowledge that everything is connected has been with Indigenous people since time immemorial. Scientists are just beginning to also declare this fact and research it and talk more about it."

FACING: Insects and other arthropods are one vector for transporting marine nutrients into terrestrial ecosystems. Top left, a snail-eating beetle (genus *Cychrus*) uses its large mandibles to pry open snail shells. Top right, a researcher with the contents of a buried pitfall trap. Center left, a sarcophagid fly. Center right, a pictured rove beetle, a predator on beach flea amphipods. Bottom left, a centipede. Bottom right, a mossy sallow moth.

SHIFTING SANDS

For kids, moving beach sand around is an abiding pleasure: buckets and shovels, sandcastles and moats, burying friends in mounds. What we don't realize as children—and something about which most adults are only vaguely aware—is that humans are not the only forces in the sand-moving game. Natural processes on dune-covered beaches can alter the landscape on a monumental scale, sometimes within the space of a single tide cycle.

The Hakai Institute Coastal Sand Ecosystems (CSE) team began researching beach erosion dynamics on Calvert Island in 2014. The CSE was founded by Ian Walker, a geomorphologist now with Arizona State University; research was assisted by a team that included geomorphologist Bernard Bauer and Michael Grilliot, then a PhD student at the University of Victoria. They found that between October 2015 and March 2016, winter storms robbed the beach of approximately 23,000 metric tonnes of sand—the weight of over 130 Boeing 747 passenger jets.

Derek Heathfield, one of the CSE team's coastal geomorphologists and mapping specialists, got a sense of the forces that alter beaches as he sat on the edge of a bluff farther south in the Pacific Rim National Park Reserve. "I was looking down at these huge logs floating and banging into one another in the surf. And then a wave would come in and just throw them at the bluff below. I was thinking that if I was down there, I'd probably be shattered into a thousand pieces. That scene had a real impact on me in terms of understanding just part of the magnitude of the forces at work in coastal systems."

To track the changes to West Beach, CSE scientists use time-lapse cameras, drones, and laser

FACING: Top, beaches on the north end of Calvert Island. Bottom, storm surge on the island's West Beach.

ABOVE: Dunegrass on Calvert Island.

scanners to create a 3D map of the area. One piece of particularly impressive technology is a U-shaped, ground-based laser particle counter they call a Wenglor, after its manufacturer, that can track the movement of hundreds of thousands of sand grains. "Instruments like these are often used in factories to count individual pieces of cereal or pills, things like that," explains Heathfield. "We set several up along the beach in a line consistent with the wind direction. It's like a game of croquet for sand. The Wenglors will actually count the individual sand grains as they move."

While dunes on beaches can lose massive amounts of sand, they can also gain it back. "The rebuilding phase of a dune follows somewhat predictable cycles or phases," explains Heathfield. One of the first visible stages of rebuilding is when you see a zone seaward of a dune where grasses start to pop up, he says. "What those grasses do is start to create roughness on the beach. If windblown sand is blowing across the beach and it encounters those grasses, then the resulting turbulence drops the sand out of transport and it collects there. That's how the dune continues to build."

Walker notes that the West Beach dunes are typical of many dune systems along the BC coast: small dune complexes trapped in rocky bays. "The dunes there are a small gem, but it's still unclear as to how representative they are of broader coastal dune resilience in other areas," he says. "The more open, dynamic coastal dune ecosystems found in Northern California, Oregon, and Washington are less common in BC, although we have some very special examples at Wickaninnish near Ucluelet or those on the Naikoon Peninsula in Haida Gwaii."

At one point, Walker was monitoring a network of 30 coastal sites from southern Vancouver Island to the northern tip of Haida Gwaii, a broad research program that he says was "a bellwether for sea-level rise and climatic variability impacts" on the northeast Pacific coast. That research needs further funding to continue, but Walker is hopeful it can be revived.

The reasons for caring about and seeking to understand the life cycle of dunes go beyond the maintenance of recreational beaches. Dunes provide crucial environmental services. Not only do they offer a buffer against large storms and accompanying sea-level rise—taking the hit, as it were, from waves that would otherwise chew away directly at the land beyond them—they are stopover points for

migratory birds. They are also primary habitat for a number of species that actually rely on the instability of the dune environment in order to thrive. Beach pea (*Lathyrus japonicus*), for instance, can only germinate when repeated dashings by waves against the sand have worn out its sturdy seed coat. "A lot of plant species thrive on burial by sand," says Heathfield, "and by being beaten up and slapped around by wind and sand. If there wasn't disturbance that kept the dune active, shore pine and other big woody species would likely totally dominate."

"Dunes change their form every tidal cycle," Walker pointed out in a Hakai Institute mini documentary. "Every season the dunes are different, which for some can pose problems. When they start to migrate or erode, we tend to get concerned. But those changes, that's part of the natural function of dunes. If left to nature's processes, they tend to recover, and they're actually quite resilient."

In short, disturbance in the dune ecosystem—erosion by winter storms or other natural

phenomena followed by a period of regrowth—promotes biodiversity and ecosystem resilience. Unfortunately, human residents of the dune ecosystem don't always appreciate the value of disturbance, and attempt to tame the shifting sands with landscaping and other interventions. As an example, Heathfield mentions the planting of grass types that prevent the movement of sand and inhibit "response and regrow following an erosive event." In the short term, says Heathfield, you may manage to protect your beachfront property from windblown sand. Over time, however—following successive heavy storms and high water levels—sand may not be available to move around and help recover the beach's natural state.

Of greater concern is the potential effect of climate change on the ability of dunes to rebuild. "What we're looking at are the factors that really drive dune erosion—the winds, the waves, the tides—and how those are linked to climate variability signals and climate change," says Heathfield. "What we know is that the frequency and the magnitude of erosive events are increasing."

While there may be a tipping point between healthy and unhealthy levels of disturbance, Heathfield isn't sure that point has been reached. He does believe, however, that dune ecosystems' ability to recover "is going to be tested" by the effects of global warming. ■

FACING: Beach pea seeds can survive for years in seawater and don't germinate until their husks are worn down by shoreline waves.

ABOVE: Dune ecosystems thrive in a cycle of erosion and regrowth.

CONCLUSION
LEARNING FROM KOEYE

CHRISTINA MUNCK Cofounder and Vice-President, Tula Foundation and Hakai Institute

IN LATE AUGUST 2005, I stepped off the dock in Bella Bella onto a fishing boat owned by Harvey Humchitt (Wiqvilba Wakas), a hereditary chief of the Heiltsuk First Nation. We chugged for three hours south down the channels to the Koeye (pronounced "kway") River estuary. It was years before we established our research center on Calvert Island, and my first time ever on the Central Coast. I was stunned by the beauty of the area.

Like Calvert Island, the Koeye watershed has been occupied by humans for many millennia. It remains one of the jewels of this coast. An 18,000-hectare watershed that was never subject to industrial-scale logging, it still has robust runs of all species of salmon, along with grizzlies, black bears, wolves, and cougars.

That year, Koeye was the site for an ongoing summer camp for First Nations youth organized by the Qqs Projects Society, a nonprofit based in Bella Bella. Qqs focuses on conservation, education, cultural revitalization, and reconnection of youth to traditional territory—all things Eric and I fully support. After visiting the site that day, I agreed that we would fund the Nature Conservancy of Canada's purchase of 67 hectares of private forestry land to add to the Koeye camp.

Calvert Island is about 20 kilometers across the channel from Koeye. Years later, in 2010, when we began to build up the Calvert Island Ecological Observatory, we immediately sought to align our efforts with those of Qqs, to learn from them, and to ensure that our programs complemented and supported theirs. We were honored at the end of that first season when the Koeye campers paddled their big canoe the many kilometers across the water from Koeye to visit us.

The work that the Hakai Institute does on Calvert Island and Quadra Island, among other places, matches my lifelong interests in coastal ecosystems and conservation. I grew up close to the ocean and chose biology as a career. These days, I enjoy being close to the science and am energized by the ingenuity and spirit of collaboration I see in our science teams and visitors. I'm inspired by the people we work with: our support staff, our in-house scientists and postdocs, and our affiliates, students, and visitors from around the world.

We are committed to ensuring that our work is integrated with our neighboring communities on the Central Coast, most of which are First Nations. Of course, all the work we do from our Calvert Island base takes place on the traditional territory of a First Nation, whether Heiltsuk, Kitasoo/Xai'xais, Nuxalk, or Wuikinuxv. Our relationships with all of them are valuable and important. In the spirit of this book, which can tell only part of the whole, the story of our work at Koeye is just one of many possible stories that illustrate these relationships.

PREVIOUS SPREAD: The Koeye Big House, seen here from a helicopter, offers a space for community gatherings.

ABOVE: Harvey Humchitt (Wiqvilba Wakas), a hereditary chief of the Heiltsuk Nation.

Building the salmon weir

In late 2011, a fire destroyed the lodge at Koeye, which set back the summer cultural and science program for a few years. The new and greatly improved lodge, designed to reflect cultural practices, fully reopened in 2014. Out of the ashes arose a remarkable complex with cabins for family visits, meeting rooms, kitchens, and play areas, plus the start of a garden for vegetables and traditional plants. It is more integrated with the community, and more connected to the Internet thanks to our partnership

with Hakai Energy Solutions, who also installed a local weather station and webcam for science and safety.

While the lodge was being rebuilt, we kept working. We were asked by Larry Jorgensen, head of Qqs, to construct a bear-proof cabin about 10 kilometers upriver on Koeye Lake to support science and resource management—in particular, to monitor spawning sockeye salmon returning to the lake's tributaries. That project—which involved helicopters and weeks in the bush—was orchestrated by Jamie Cowan of Qqs, who designed the cabin, and by Hakai's resident photographer, wilderness guide, and "maker," Grant Callegari.

With the science cabin in place, the salmon-monitoring project moved to the next phase.

One of the keys to successful science in communities is collaboration with knowledgeable outsiders who are attuned to working with local experts like Qqs. Typically, the brunt of the academic scientific work is borne by students and postdocs. In the case of the Koeye salmon project, the academic partner was Simon Fraser University (SFU), and the key person in the formative years was PhD student Will Atlas.

A mainstay of salmon research is the counting fence. Typically, the fence is a utilitarian structure that spans the river near its mouth—allowing researchers to count, sample, or tag salmon heading upstream to

LEFT: Grant Callegari and Koeye crew prepping materials for the fish weir.

RIGHT: Will Atlas releasing a steelhead.

LEFT: Treehouse platform built as the summer residence for operators of the weir.

RIGHT: Richard Wilson-Hall, foreground, and Peter Taylor moving a finished weir tripod.

spawn. After discussions with Heiltsuk leaders, Grant, Will, and Larry met in Bella Bella in the spring of 2013 to explore a better idea. Weirs have traditionally been used on this coast for harvesting salmon. Why not modify this traditional design for use as a counting fence? Grant researched designs, in part from a book by Hilary Stewart, and formulated plans with the Qqs crew.

Construction and installation of the weir became a significant community project that involved the sawmill in Bella Bella, brains and brawn from the local community, and many volunteers from Hakai and elsewhere. The weir was up and running by June, allowing Heiltsuk and SFU scientists to capture a subset of the returning sockeye for tagging.

The process is a bit like customs and immigration. Selected fish are classified, measured, and fitted with visual tags as well as passive integrated transponder (PIT) tags. They are then sent on their way upriver into Koeye Lake and ultimately its spawning tributaries. Strategically placed detector arrays, supported from the science cabin, monitor their progress to the spawning sites. Along with providing valuable data on Koeye salmon, the fish weir also acts in the classic way—as a method of community food harvest, revitalizing a traditional approach to subsistence fishing.

This to us is the perfect project, combining great science with conservation, resource management, education, culture, and even art, all in one beautiful package.

TOP AND BOTTOM LEFT:
Robert Duncan helping to
construct a traditional
style of fish weir.

BOTTOM RIGHT: Grant
Callegari wading through
the Koeye estuary with
camera equipment.

FACING: The completed fish weir.

TOP: The waters of the Koeye River moving through the weir pickets.

BOTTOM: Scott Lawson retrieving sockeye from the trap box for tagging.

Grizzly bears on the Central Coast. Grizzlies were the target of a controversial trophy hunt in British Columbia that was banned in 2017.

Snagging bear hair

The salmon ecosystem at Koeye attracts that other iconic species, the grizzly bear. One of the symbols of the Great Bear Rainforest, grizzly bears are important to ecologists as top predators, and of particular cultural significance to First Nations. They are also much sought after by trophy hunters.

When we started work on the coast in the mid-2000s, there was great contention between the trophy hunters and those who wanted to stop the hunt. The provincial government was in the middle, responsible for regulating the hunt, but their regulations weren't based on sound science. Nobody really knew how many bears there were, how far they ranged, or what the population structure was. In the absence of data, regulators fell back on abstract models.

This was another critical area where scientists and First Nations could find common ground. Led by William Dúqv̓aísl̓a Housty, in 2006 Qqs staff and Heiltsuk youth set up bear-hair-snagging stations in the Koeye watershed. A snagging station is scented to attract bears. It measures about 12 square meters and is bounded by a low line of barbed wire around several trees. As the bears clamber over the wire to investigate the smell, they leave tufts of hair on the barbs. These hair samples are collected, labeled, and sent for DNA analysis—it's CSI for bears. Given enough stations, enough samples, and enough time, it is in theory possible to complete an inventory of local bears while learning about their movements, their relationships to salmon, and other aspects of their lives.

This is a nontrivial scientific project. Luckily, we benefited from the local and traditional knowledge of Qqs and a great academic collaborator, Chris Darimont, who began working on the bear project in 2009. We knew of Chris because he had previously done his PhD thesis on the wolves of Calvert Island and beyond.

Around that same time, and inspired by William's lead, neighboring Central Coast First Nations expressed interest in doing similar work. I was unreservedly enthusiastic about supporting that project and its marriage of academic science and traditional knowledge. Like most people, Eric and I loved grizzly bears at an emotional level and were strongly opposed to the trophy hunt. We distrusted the science that was then informing decision-making by the government, and were already major contributors to the Raincoast Conservation Foundation's trophy license buyback campaign. We combined efforts with Raincoast to support Chris for a five-year appointment as the Hakai–Raincoast Professor at the University of Victoria, allowing him to work with the Koeye bear monitoring project, among other things.

We particularly liked the collaboration among the four First Nations. It was rewarding to see our good friends all working together, and from a practical perspective—given that grizzlies roam over a huge range and pay no attention to boundaries between territories—the project could not be done properly without collaboration.

TOP: William Dúqvȧíṣḷa Housty, director of the Qqs Project Society's Coastwatch program, a Heiltsuk-driven scientific research initiative.

BOTTOM: Megan Adams, left, and Chris Darimont looking at bear hair on a trip wire.

Thanks to all this work, there is detailed and credible ecological knowledge of grizzlies across the Central Coast. This solid scientific data was a significant contribution—along with a related scientific audit of the hunt and a sophisticated campaign by First Nations—to the decision by the BC government to close the trophy hunt at the end of 2017. Seeing science and traditional ecological knowledge working together to inform sound policy and meet community priorities was an inspiration.

A place of learning

The collaboration at Koeye has had so many other lasting benefits. We see a new generation of confident young people who have been through the Qqs programs, who are now big contributors to their communities and to the Hakai Institute. University students have also had opportunities to gain experience in a rich scientific and cultural setting. But the flow of knowledge and mentoring is a two-way process, and Koeye has been a place of deep learning for everyone involved, including myself.

Going back to that boat ride to Koeye in 2005, my ignorance of culture and place was obvious. What an education it has been to work with all these people and see these wonderful projects unfold.

LEFT: Sockeye salmon being roasted by the fire.

RIGHT: Top, students from Bella Bella visiting the fish weir. Bottom, a grizzly bear feeding on salmon.

ACKNOWLEDGMENTS

THE AUTHOR GRATEFULLY acknowledges the following people, whose time, resources, research, and insights made this book possible:

Jeff Bond; Jenn Burt; Chris Darimont; Emmett Duffy; Christopher Ernst; James Estes; Wiley Evans; Daryl Fedje; Erin Foster; Alisha Gauvreau; Alyssa-Lois Gehman; Ian Giesbrecht; Christopher Harley; Leslie Harris; Drew Harvell; Derek Heathfield; Margot Hessing-Lewis; Brent Hughes; Brian Hunt; Jennifer Jackson; Patrick Keeling; Colleen Kellogg; Nancy Knowlton; Lynn Lee; Matt Lemay; Keith MacLachlan; Christopher Mah; Duncan McLaren; Brian Menounos; Christina Munck; Linda Nichol; Daniel Okamoto; Gustav Paulay; Eric Peterson; Ondine Pontier; Anne Salomon; Brian Starzomski; Curtis Suttle; Ian Walker; Jane Watson; Matt Whalen; and Sara Wickham.

Thanks are due also to Grant Callegari and Josh Silberg for their tremendous assistance with photography and species identification; to Jacqueline Moore and Paul Papin for their skilled help with research and editorial; to Chris Labonté at Figure 1 Publishing for facilitating the early discussions; to the Figure 1 production team, in particular Naomi MacDougall, Jess Sullivan, and Lara Smith; to Richard Nadeau and Mark Redmayne in sales and marketing at Figure 1; and to the sales and marketing teams at Raincoast Books and Publishers Group West.

FURTHER READING

The Peopling of the Americas

"A Revised Sea Level History for the Northern Strait of Georgia, British Columbia, Canada," *Quaternary Science Reviews* 192 (2018). Daryl Fedje, Duncan McLaren, Thomas S. James, Quentin Mackie, Nicole F. Smith, John R. Southon, and Alexander P. Mackie

"Fladmark + 40: What Have We Learned about a Potential Pacific Coast Peopling of the Americas?" *American Antiquity* 85 (2020). Todd J. Braje, Jon M. Erlandson, Torben C. Rick, Loren Davis, Tom Dillehay, Daryl W. Fedje, Duane Froese, Amy Gusick, Quentin Mackie, Duncan McLaren, Bonnie Pitblado, Jennifer Raff, Leslie Reeder-Myers, and Michael R. Waters

Haida Gwaii: Human History and Environment from the Time of Loon to the Time of the Iron People. UBC Press, 2005. Daryl W. Fedje and Rolf W. Mathewes

"Routes: Alternate Migration Corridors for Early Man in North America," *American Antiquity* 44 (1979). Knut R. Fladmark

"Terminal Pleistocene Epoch Human Footprints from the Pacific Coast of Canada," *PLOS ONE* 13 (2018). Duncan McLaren, Daryl Fedje, Angela Dyck, Quentin Mackie, Alisha Gauvreau, and Jenny Cohen

Glaciers and Coastal Ecosystems

"Glacier Retreat and Pacific Salmon," *BioScience* 70 (2020). Kara J. Pitman, Jonathan W. Moore, Matthew R. Sloat, Anne H. Beaudreau, Allison L. Bidlack, Richard E. Brenner, Eran W. Hood, George R. Pess, Nathan J. Mantua, Alexander M. Milner, Valentina Radić, Gordon H. Reeves, Daniel E. Schindler, and Diane C. Whited

"Glacier Shrinkage Driving Global Changes in Downstream Systems," *Proceedings of the National Academy of Sciences* 114 (2017). A.M. Milner, K. Khamis, T.J. Battin, J.E. Brittain, N.E. Barrand, L. Füreder, S. Cauvy-Fraunié, G.M. Gíslason, D. Jacobsen, D.M. Hannah, A.J. Hodson, E. Hood, V. Lencioni, J.S. Ólafsson, C.T. Robinson, M. Tranter, and L.E. Brown

"Icefield-to-Ocean Linkages across the Northern Pacific Coastal Temperate Rainforest Ecosystem," *Bioscience* 65 (2015). S. O'Neel, E. Hood, A.L. Bidlack, S.W. Fleming, M.L. Arimitsu, A. Arendt, E. Burgess, C.J. Sergeant, A.H. Beaudreau, K. Timm, G.D. Hayward, J.H. Reynolds, and S. Pyare

"When the Glaciers Disappear, Those Species Will Become Extinct," *New York Times*, 2019, https://www.nytimes.com/interactive/2019/04/16/climate/glaciers-melting-alaska-washington.html. Henry Fountain

Marine Food Webs

A Guide to the Marine Plankton of Southern California. 3rd ed. UCLA OceanGlobe, 2003. Robert Perry

Biological Oceanography: An Introduction. 2nd ed. Butterworth-Heinemann, 1997. Carol M. Lalli and Timothy R. Parsons

"Marine Microbes," Smithsonian Ocean, https://ocean.si.edu/ocean-life/microbes/marine-microbes. Danielle Hall

Microbial Ecology of the Oceans. 3rd ed. Wiley-Blackwell, 2018. Josep M. Gasol and David L. Kirchman

Plankton: A Guide to Their Ecology and Monitoring for Water Quality. 2nd ed. CRC Press, 2019. Iain M. Suthers, David Rissik, and Anthony J. Richardson, eds.

"Special Issue on A Sea of Microbes," *Oceanography* 20 (2007).

Viruses and Infectious Disease

Ocean Outbreak: Confronting the Rising Tide of Marine Disease. University of California Press, 2019. Drew Harvell

Spillover: Animal Infections and the Next Human Pandemic. W.W. Norton, 2012. David Quammen

The Malaria Capers: More Tales of Parasites and People, Research and Reality. W.W. Norton, 1991. Robert S. Desowitz

Kelp Forests, Urchins, and Sea Otters

Otter Skins, Boston Ships, and China Goods: The Maritime Fur Trade of the Northwest Coast, 1785–1841. McGill-Queen's University Press, 1992. James Gibson

"Restoring the Kelp Forests of Gwaii Haanas," Gwaii Haanas National Park Reserve, National Marine Conservation Area Reserve, and Haida Heritage Site, https://www.pc.gc.ca/en/pn-np/bc/gwaiihaanas/nature/conservation/restauration-restoration/nurture-nourrir-1.

Return of the Sea Otter: The Story of the Animal That Evaded Extinction on the Pacific Coast. Sasquatch Books, 2018. Todd McLeish

Sea Urchins: Biology and Ecology. 4th ed. Academic Press, 2020. John M. Lawrence

Serendipity: An Ecologist's Quest to Understand Nature. University of California Press, 2016. James A. Estes

"Smashing Urchins for Kelp," *Hakai Magazine*, 2019, https://www.hakaimagazine.com/videos-visuals/smashing-urchins-for-kelp/. Katrina Pyne, Grant Callegari, and Jude Isabella

Seagrass

"A Call for Seagrass Protection," *Science* 361 (2018). Leanne C. Cullen-Unsworth and Richard Unsworth

Project Seagrass, https://project-seagrass.webs.com/.

World Atlas of Seagrasses. University of California Press, 2003. Edmund P. Green and Frederick T. Short

Oceanography

"Basin-Scale Oceanographic Processes, Zooplankton Community Structure, and Diet and Reproduction of a Sentinel North Pacific Seabird over a 22-year Period," *Progress in Oceanography* 182 (2020). J. Mark Hipfnera, Moira Galbraith, Douglas F. Bertram, and David J. Green

"Biological Impacts of the 2013–2015 Warm-Water Anomaly in the Northeast Pacific: Winners, Losers, and the Future," *Oceanography* 29 (2016). Letícia M. Cavole, Alyssa M. Demko, Rachel E. Diner, Ashlyn Giddings, Irina Koester, Camille M.L.S. Pagniello, May-Linn Paulsen, Arturo Ramirez-Valdez, Sarah M. Schwenck, Nicole K. Yen, Michelle E. Zill, and Peter J.S. Franks

"Huge 'Hot Blob' in Pacific Ocean Killed Nearly a Million Seabirds," *Guardian*, 2020, https://www.theguardian.com/environment/2020/jan/16/hot-blob-ocean-seabirds-killed-new-zealand-north-america. Kenya Evelyn

Oceanography of the British Columbia Coast. Gordon Soules Book Publishers, 1981. Richard Thompson

"Special Issue on Ocean Warming," *Oceanography* 31 (2018).

Marine Biodiversity

Citizens of the Sea: Wondrous Creatures from the Census of Marine Life. National Geographic, 2010. Nancy Knowlton

Log from the Sea of Cortez. Penguin Classics, 1941. John Steinbeck

Mac's Field Guides: Northwest Coastal Invertebrate (laminated guide). The Mountaineers Books, 1988. Craig MacGowan

Marine Life of the Pacific Northwest: A Photographic Encyclopedia of Invertebrates, Seaweeds and Selected Fishes. Harbour Publishing, 2006. Andy Lamb and Bernard Hanby

Seashore Life of the Northern Pacific Coast: An Illustrated Guide to Northern California, Oregon, Washington, and British Columbia. University of Washington Press, 1983. Eugene N. Kozloff

The Intertidal Wilderness: A Photographic Journey through Pacific Coast Tidepools. Rev. ed. University of California Press, 2002. Anne Wertheim Rosenfeld and Robert T. Paine

The New Beachcomber's Guide to the Pacific Northwest. 3rd ed. Harbour Publishing, 2019. J. Duane Sept

General Biodiversity

"Grizzly Bear Monitoring by the Heiltsuk People as a Scientific Crucible for First Nations Conservation Practice," *Ecology and Society* 19 (2014). W.G. Housty, A. Noson, G.W. Scoville, J. Boulanger, C.T. Darimont, and C.E. Filardi

Staying the Course, Staying Alive: Coastal First Nations Fundamental Truths: Biodiversity, Stewardship and Sustainability. Biodiversity BC, 2009. Frank Brown and Y. Kathy Brown, comps.

The Diversity of Life. 2nd ed. Harvard University Press, 2010. E.O. Wilson

The Invention of Nature: Alexander von Humboldt's New World. Vintage, 2016. Andrea Wulf

The Last Wild Wolves: Ghosts of the Great Bear Rainforest. Greystone Books, 2007. Ian McAllister, Paul Paquet, and Chris Darimont

The Sixth Extinction: An Unnatural History. Picador, 2015. Elizabeth Kolbert

Living Shorelines

Guidance for Considering the Use of Living Shorelines. National Oceanic and Atmospheric Administration, 2015.

"San Francisco Bay Living Shorelines Project," http://www.sfbaylivingshorelines.org.

Ocean Optimism

A World for My Daughter: An Ecologist's Search for Optimism. Caitlin Press, 2015. Alejandro Frid

Changing Tides: An Ecologist's Journey to Make Peace with the Anthropocene. Caitlin Press, 2015. Alejandro Frid

Ocean Optimism, http://www.oceanoptimism.org/.

INDEX

Page numbers in italics indicate an image; page numbers in bold indicate a map.

5, 6, 43–45, *50*, 50–53, *51,*
52, 53, 115, 117, 118–20, 120,
124–25 (*see also* Hunt, Brian;
Jackson, Jennifer); 100 Islands
project, 141–44; overview, vi,
viii–ix; salmon, *5*, *6*, 40, 43,
45, 52–53, *53*, 119, *121*, *159*,
159–60, *161–63*; sea otters,
77–78, 85, 91; sea stars, 64,
72, *73*, 74; sea urchins, 96–97;
watershed science program,
34 (*see also* Giesbrecht, Ian).
See also Airborne Coastal
Observatory; Calvert Island
Ecological Observatory (BC);
Hakai Ancient Landscapes
Archaeology Project; Hakai
Cryosphere Node; Marna Lab
(BC); Quadra Island Ecological
Observatory (BC)
Hakai Magazine, vi, 2
Hakai Passage (BC), 16, 17
Hakai Spirit, 52
Hall, Kyle, *67*
harbor seals (*Phoca vitulina*),
39, *105*
Harley, Christopher, 61–62,
62, 64, *72*, 74
Harris, Leslie, 131
Harvell, Drew, 67, *70*, 70–71,
72, 74, 107
Heathfield, Derek, 153–54,
155
Heiltsuk First Nation:
archaeological projects, 12;
Coastwatch program, 165;
collaboration with Hakai
Institute, 4–5, 158; grizzly
bear project, 164–66; Koeye
salmon weir, 160, *161–63*;
traditional territory, 15,
158; Triquet Island, 22–23.
See also Humchitt, Harvey
(Wiqvilba Wakas)
herring, 43, 78, *104*
Hessing-Lewis, Margot, 110
Heterosigma akashiwo,
54–55
Hewson, Ian, 71
holdfasts, 78, 103, *103*
Holmes, Keith, 89
Homathko weather station
(BC), *28*
Honka, Leah, 91
Housty, William Dúqváísla, 22,
164–65, *165*
Hughes, Brent, 108, 109

Humchitt, Harvey (Wiqvilba
Wakas), 157, *158*
Hunt, Brian, 43–45, 46–47, *51,*
52–53, 113–15, 118–20

ice ages, 11, 15–16, 17–19, **18**.
See also Beringia; glaciers
insects, terrestrial, *135, 137,*
146, *150*
International Union for
Conservation of Nature, 93
intertidal zone, 58, 62
invertebrates, microbial, and
glacier meltwater, 33
Isabella, Jude, vi
isopods, 108, *132*

Jackson, Jennifer, 115–16,
117–18, 121
Japan, 95, 96
jellyfish, *xii*, 45, *46*, 47, *48*
Jo, Sung, *135*
Joffre Lakes (BC), 33
Johnson, Brett, *53*
Johnson, Patrick, *166*
Johnstone Strait (BC), *121*
Jorgensen, Larry, *159*,
160

Kay, Sharon, 74
Kellogg, Colleen, 35, 43, 48,
50, 50–51
kelp: about, 78, 103; in Central
Coast food webs, 143–45;
climate change, 120; dumps
carbon into nearshore
ecosystem, 79; in marine food
webs, 78–79. *See also* bull
kelp (*Nereocystis luetkeana*);
seaweed; sea wrack
kelp forests: about, 78–79, 81;
assets to human coastal
communities, 81; biodiversity,
74, 77–78; on Central Coast,
77–78, 144; impact on
nearshore habitat, 79, 81;
mapping, 89; ocean warming,
88; research, 77–78, *78*, 85,
87–89, *89*; rockfish, *76–77,*
81; salmon, 78; seals, *105*;
sea otters, 77–78, 82–83,
84, 85, 87–88, 97; sea stars,
78–79, 88; sea star wasting
disease, 74, 87–88; sea
urchins, 70, 82–83, *83*, 85,
88, 95
kelp greenling, *101*

kelp highway, **18**, 19, 21, 81.
See also North America:
human migration routes
Kenilworth Park and Aquatic
Gardens (DC), 128
keystone species: biodiversity,
57–60, 74; definition, 59–60,
102, *103*; pathogen, 104.
See also sea otters; sea stars;
trophic cascade
Kilbella Bay (BC), *10–11*
killer whales, 78, 92
Kîsik Aerial Survey, 30
Kitasoo/Xai'xais First Nation,
4, *158*, 165–66
Klinaklini Glacier (BC), 29, *31, 32*
Klinaklini River (BC), *32*, 33, *37*
Klinaklini weather station
(BC), *36*
Knight Inlet (BC), 33, *61*
Knowlton, Nancy, viii–ix
Koeye (BC), **x**, 157–60, 164–65,
166
Koeye Big House, *156–57*
krill, 46, *49*

Labyrinthula zosterae
(Labby), 107
Lawson, Scott, *163*
leather stars (*Dermasterias*
imbricata), 74, *75*
Lee, Lynn, 95–96, *96*, 97
Lemay, Matt, 134–35
Lertzman, Ken, 5–6
Limpet, the, *120*
limpets, 58
Long Term Ecological Research
(LTER) Network, 5

Makah Bay (WA), 57, 58–59,
60, *60*
Makah Indian Tribe, 57
mammoth, woolly, 12
marine food webs: carbon, 48,
53; climate change, 50–51,
52–53 (*see also* Blob, the);
disease, 51, 52; in Elkhorn
Slough, 108–9; kelp, 78–79;
plankton, 40, 43–45, *44*, *45*,
47, 47–48, *48–49*; research,
43–45; prey quality, 52;
research, 47, 50–53, 118–20;
rivers and streams, 33–35; in
Salish Sea, *44*; salmon, 39–40,
43, *44*, 46–47, 52–53, 117,
118; seagrasses, 101–2, 107,
108–9; sea stars, 51, 58, 66;

sea urchins, 78–79; viruses,
40, 43–44, *44*, 46–47, 48, 52,
54–55, *55*. *See also* keystone
species
Marine Global Earth Observatory
(MarineGEO), 99, 127, 130, 134
marine heat waves. *See* climate
change
marine subsidies, 140–42, 148.
See also food webs; marine
food webs
Marna Lab (BC), *6, 50*. *See also*
Quadra Island Ecological
Observatory (BC)
Mauna Loa Observatory (HI), 125
McLaren, Duncan, 12, *13, 14,*
16–17, 19–20, *20*, 21–22
Menounos, Brian, 25, 29, 30,
32–33, 37
mink, 148
Mitra, 2, 3, 8
mola mola, *115*
Monterey Bay (CA), 3–4, 86, 108
Monterey Bay Aquarium Research
Institute (CA), 3–4
Monte Verde (Chile), 19
morning sun stars (*Solaster*
dawsoni), 61, 66
mottled stars (*Evasterias*
troschelii), 61, 74
Mount Jacobsen (BC), *24–25*
Mount Waddington (BC), *27*, 29
Multi-Agency Rocky Intertidal
Network (MARINe), 5
Munck, Christina, vi, 1, 2–3, *4*, *158*
mussels, 58–59, *59, 62*, 67, 91, 125

National Marine Conservation
Area Reserve (BC), 96
National Oceanic and
Atmospheric Administration
(NOAA) (US), 124–25
National Science Foundation
(US), 5, 70
Natural History Museum of
Los Angeles County (CA), 131
Nelson, Trisalyn, 142
Newton, Harley, 71
Nichol, Linda, 89, 91
Niskin bottles, 50, 123
North America: early human
occupation of, 11–12, 15, 16–23;
human migration routes, 11, 17,
18, 19, 20–22; during last ice
age, 11, 15, 17, **18**
nudibranchs, *81, 106, 133*
Nuu-chah-nulth Tribal Council, 93

60, 83; maritime fur trade, 92; ocean warming/the Blob, 72; reestablishment in northeast Pacific, 60, 77, 87, 92, 97; research, 77–78, 82–83, 85, 87–89, 90, 91; seagrass, 108–9, 110–11; sea stars, 66; sea urchins, 82–83, 88, 91; urchin barrens, 96, 97

sea slugs, 49, 81, 108–9, 109. *See also* nudibranchs

sea snails, 58, 62, 67, 91, 129, 130. *See also* abalone

sea stars: about, 63, 64–67, 66, 67; climate change, 72, 74; detaching arms, 60, 66; impact on Central Coast ecosystem, 87–89; indicate ecosystem health, 62; in kelp forests, 78–79, 88; larvae, 79, 81; in marine food webs, 58, 66; misidentified, 133; in northeast Pacific Ocean, 65, 66, 68–69; research, 61–62, 64, 71–72, 73, 74. *See also* ochre stars; sea star wasting disease; sunflower stars

sea star wasting disease (SSWD): description, 60–61, 64, 75; early outbreaks, 60–62, 64, 71; impact on Central Coast ecosystem, 87–89; kelp forests, 74, 87–88; "landmark event," 70; marine food webs, 51; ocean warming, 71–72, 74; research, 71–72, 73, 74; sea urchins, 70, 88, 95

sea urchins: about, 65; barrens, 83, 83, 94, 95–97; in British Columbia, 95, 96–97; in California, 95, 96; in Caribbean, 70; commercial harvest, 95, 97; kelp forests, 70, 82–83, 83, 85, 88, 95; in marine food webs, 67, 78–79; in marine subsidies, 140; ocean warming, 95; in Oregon, 95; research, 89, 95–97, 96; roe, 91, 96, 97; sea otters, 82–83, 88, 91; sea star wasting disease, 70, 88, 95; underside view, 94; zombies, 95–97

seaweed: about, 103, 103; sea stars, 58. *See also* kelp

sea wrack, 143, 144–45, 145, 146, 146, 147, 148, 151

Shark Bay (Australia), 102

sharks, 39, 92, 102, 109, 110, 115

shellfish, 21, 93, 125. *See also* abalone; clams; crabs; mussels; oysters; sea urchins

Shemya Island (AK), 83

Silverthrone Mountain (BC), 29

Simon Fraser University (BC), 5, 141, 159

Smith, Fred, 57–58

Smith, John Maynard, 9

Smithsonian Institution (US), 5, 8, 99, 107, 130. *See also* Marine Global Earth Observatory (MarineGEO)

songbirds, 60, 146, 148

Sonoma State University (CA), 108

spiders, 137

Spindel, Nathan, 97

spiny lumpsucker, 131

sponges, 58

squid, 148

Starzomski, Brian, 139, 141, 142, 148

Staude, Craig, 131

Stewart, Hilary, 160

St. Pierre, Kyra, 35

Strait of Georgia/Salish Sea, 6, 40, 43, 44, 61, 113, 114

striped seaperch, 102

striped sun stars (*Solaster simpsoni*), 69, 75

sunfish, 115

sunflower stars (*Pycnopodia helianthoides*): hunting and eating techniques, 67; kelp forests, 79, 88; keystone species, 70; in northeast Pacific, 61, 65, 68, 70; predator, 66; sea urchins, 78–79, 88, 95; wasting disease, 60–61, 64, 64, 70, 72, 74

sun stars, 75

surfgrasses, 98–99, 105. *See also* seagrasses

sushi, 95, 96

Suttle, Curtis, 54–55

Tatoosh Island (WA), 5, 59

Taylor, Peter, 160

transects, 73, 85

Triquet Island (BC), **18,** 21, 22–23

trophic cascade, 60, 83, 108–9

Tula Foundation (BC), vi, 2–3. *See also* Hakai Institute (BC)

University of British Columbia (BC), 8, 54, 61, 123

University of California Davis (CA), 107

University of Central Florida (FL), 107

University of Northern British Columbia (BC), 8, 25, 30. *See also* Hakai Cryosphere Node

University of Victoria (BC), 2, 22, 141, 142, 153, 165

University of Washington (WA), 57, 125, 131

urchin barrens, 83, 83, 94, 95–97

Vancouver (BC): map, **xi**; sea stars, 61–62, 74. *See also* Simon Fraser University (BC); University of British Columbia (BC)

Vancouver Island (BC): map, **xi**; sea otters, 92; sea star wasting disease, 64. *See also* Campbell River (BC); Mount Jacobsen (BC); Pacific Rim National Park Reserve (BC); Victoria (BC); Yellow Point (BC)

Vancouver Island University, 25, 26, 28, 30, 36

Victoria (BC), **xi,** 8

viruses: as infectious pathogens, 71; in marine food webs, 40, 43–44, 44, 46–47, 48, 52, 54–55, 55

Walker, Ian, 153, 154

Walkus, Jennifer, 129

Washington (US), 5, 26, 58, 60, 65, 70, 107, 154. *See also* University of Washington (WA)

watercraft, ancient, 19, 23, 23

Watson, Jane, 77, 79, 85, 89, 91, 111

Wenglor laser particle counter, 153

Whalen, Matt, 127, 131, 132, 136

whales: and barnacles, 149; on Central Coast, 127; fin, 72; gray, 78; humpback, 149; kelp forests, 79; and wolves, 148

Wickham, Sara, 143, 144–46, 145, 151

Wilson-Hall, Richard, 160

wolves, gray (*Canis lupus*): on Calvert Island, 165; diet, 118, 139, 148; on Goose Islands, 7, 139–41; in Greater Yellowstone Ecosystem, 60; on islands, 138–39, 141, 151; in Koeye watershed, 157; "salmon forest," 140

worms, 49, 131, 132, 133

Wuikinuxv First Nation: archaeological projects, 12; collaboration with Hakai Institute, 4, 6, 158, 165–66; in Rivers Inlet, 6, 117–18; traditional territory, 4, 15, 158

Yellow Point (BC), 64

zombie urchins, 95–97

zooplankton, 43–45, 44, 46–47, 48, 53. *See also* plankton

Zostera Experimental Network, 99, 102